Weak and variational methods for moving boundary problems

C M Elliott & J R Ockendon

Weak and variational methods for moving boundary problems

Pitman Advanced Publishing Program
BOSTON · LONDON · MELBOURNE

PITMAN BOOKS LIMITED
128 Long Acre, London WC2E 9AN

PITMAN PUBLISHING INC
1020 Plain Street, Marshfield, Massachusetts

Associated Companies
Pitman Publishing Pty Ltd, Melbourne
Pitman Publishing New Zealand Ltd, Wellington
Copp Clark Pitman, Toronto

AMS Subject Classifications: (main) 35-XX, 65-XX, 80-XX
(subsidiary) 34-XX, 49-XX, 76-XX

Library of Congress Cataloging in Publication Data

Elliott, C.M.
Weak and variational methods for moving boundary problems.—(Research notes on mathematics; 59)
Bibliography; p.
Includes index
1. Boundary value problems
I. Ockendon, J.R. II. Title. III. Series
QA379.E43 515.3′5 81-21136
AACR2

British Library Cataloguing in Publication Data

Elliott, C.M.
Weak and variational methods for moving boundary problems.—(Research notes on mathematics; 59)
1. Boundary value problems
I. Title II. Ockendon, J.R. III. Series
515.3′5 QA379

ISBN 0-273-08503-4

Reproduced and printed by photolithography
in Great Britain by Biddles Ltd, Guildford

Preface

This book owes its origins to a scheme developed at the Mathematical Institute in Oxford to encourage the application of mathematics to problems occurring in the real world, in particular in industry. The realisation that perhaps a quarter of the differential equations problems arising in this scheme were to be solved in a region or domain which was not known in advance came as a surprise to everyone involved in view of the general lack of coverage at student level of these "free and moving boundary problems." In the past two decades there has, however, been a burgeoning of the mathematical literature on the subject and our principal aim in writing this book is to distil and explain what seem to be the most useful parts of the theory from the practical point of view.

We should not have been able to do this without a great deal of help from our colleagues who have been involved with the Oxford scheme, both locally and as visitors, and who are too numerous to mention here. However, the scheme itself would never have existed but for the efforts of Dr A. B. Tayler and Prof L. Fox and this book would never have existed without Mrs M. Phipps' patience at the typewriter. We are grateful to them all.

CME

JRO Sept. 1981

Contents

Introduction

Many problems in various areas of applied science can be modelled as ordinary or partial differential equations posed in domains whose boundaries are to be determined as part of the problem. Depending on whether these boundaries are stationary or moving, such problems are usually referred to as free or moving boundary problems, despite possible confusion with stress-free boundaries in continuum mechanics.

Free and moving boundary problems are, in general, even harder to solve, either analytically or numerically, than the underlying differential equations would be in a prescribed domain. In particular, two separate solutions of such problems cannot be superposed, and this inherent nonlinearity means that there is a dearth of explicit analytical solutions. Indeed, almost all the solutions to be found in the literature are of a numerical nature and these solutions can exhibit a great variety of different types of behaviour. For example, moving boundaries can suddenly vanish or appear, or move with infinite speed or become blurred, all with concomitant singularities in the differential equation. Even in the absence of singularities, the example of dendritic crystal growth illustrates how complicated moving boundaries can become geometrically.

The theme of this book will be the analysis of a fairly small class of free and moving boundary problems which have a special unifying mathematical structure. This structure results from the equivalence of their classical formulations, as differential equations with suitable explicit conditions at the prescribed and unknown boundaries, to either a weak formulation in terms of integral identities or a variational statement. In neither of these generalised formulations does the free or moving boundary appear explicitly, and they are therefore referred to as "fixed domain" formulations. They enable the analysis of existence and uniqueness of solutions to proceed with the minimum of interference from the unknown boundary as well as suggesting numerical algorithms for solving the problems without explicitly tracking the position of the unknown boundary. However, against these advantages must be set the relative obscurity of the qualitative properties of the solution of the weak or variational formulation. In particular, asymptotic analyses which can greatly enhance the intuitive understanding of certain problems are much more readily carried out on the classical formulation.

Our intention in writing this book is to present these ideas, in as accessible a form as possible, to applied research workers who may be confronted with unfamiliar free or moving boundary problems and be discouraged by the diversity and abstract nature of known mathematical results about such problems. In these situations, when an explicit analytical solution is unlikely to be available, it will usually be possible to devise a numerical method which gives physically sensible results. However, to justify such a method may be difficult and we therefore consider it worthwhile to draw attention to the ease of implementing and justifying algorithms suggested by weak and variational methods, albeit for a restricted class of problems.

Our layout will be to begin in Chapter I with descriptions of a variety of situations which give rise to free and moving boundary problems. Although some of these situations do not suggest weak or variational reformulations in any obvious way, we hope that brief mention of a broad class of problems will put such reformulations into perspective. Then in Chapter II we will give an intuitive account of the reformulation of those models which are suitable. In

Chapters III and IV we will outline how, for some of the simpler examples of Chapter I weak and variational formulations can be rigorously related to classical ones and in these chapters we will also discuss the existence, uniqueness and smoothness of solutions as well as the validity and usage of some "fixed domain" algorithms. In Chapter V we will briefly reconsider the remaining examples of Chapter I, only giving references to proven results, but some of these results will enable us to speculate about the physical applicability of generalised solutions. Finally, in Chapter VI, we will review the theory we have presented and try to indicate the most general types of problem to which it can be applied.

Throughout the book, we will be exclusively concerned with direct free and moving boundary problems as distinct from those of a control theory nature, where the boundary is encouraged to have certain properties by tuning the data in the underlying differential equations or on the fixed boundaries. Also we will not attempt to trace the historical development of weak and variational methods.

For those interested in a more general and thorough account of the mathematical methods available for free and moving boundary problems, there are disappointingly few reference works relative to the large number of published research papers. Indeed, there is none which describes either a broad spectrum of problems or of methods for their solution. There are a handful of books devoted to different theoretical aspects [10, 70, 91, 140] six conference proceedings [3, 63, 105, 157, 180, 249] and a few review articles [80, 138, 163, 168, 195, 210]. Also, the breadth of the subject is indicated in the useful bibliographies [82, 175].

I The Origins of Free and Moving Boundary Problems

We will be primarily interested in ordinary or partial differential equation models of physical situations in which rapid changes occur across a region which is thin enough to be modelled as a sharp boundary. The crucial feature of the models which concern us is that the position of this boundary is to be determined as part of the solution to the problem. Such situations arise in many models in physical and nonphysical applied mathematics and in this chapter we will try to mention briefly a spectrum of such problems for which there are reasonable mathematical theories, not necessarily interrelated. Thereby we will try to put into perspective those few models for which unifying reformulations exist.

The descriptions given below are those we feel describe most simply and directly the relevant physical situation. All questions concerning different types of reformulation and their comparative assessment will be deferred to Chapter II.

1. FREE AND MOVING BOUNDARY PROBLEMS INDUCED BY THE PHYSICAL INACCEPTABILITY OF SOLUTIONS OF THE UNDERLYING DIFFERENTIAL EQUATION: SHOCK WAVES

This is an intensively studied class of free and moving boundary problems with a history at least as long as that of gas dynamics [67]. The prototype situation concerns the solution of a nonlinear hyperbolic equation such as

$$\frac{\partial u}{\partial t} + u\,\frac{\partial u}{\partial x} = 0, \quad u(x,0) = u_0(x); \tag{1.1}$$

here u might represent the velocity of a gas in a tube $-\infty < x < \infty$. Even for analytic data, say $u_0 = \sin x$, it is easy to see that u will in general steepen before becoming multivalued after a finite time $t = t_{crit}$. For a physically acceptable solution for times $t \geq t_{crit}$, our model must be modified to meet the physical requirement of single-valuedness of u. This can only be done at the expense of introducing an unknown boundary, called a shock wave in this case, across which u can have a "step" or "jump" discontinuity. This shock wave can be introduced in an infinite number of different ways unless we add further requirements from physical considerations. The foremost of these arises when (1.1) is derived from a conservation law of continuum mechanics in the form

$$\frac{\partial P}{\partial t} + \frac{\partial Q}{\partial x} = 0 \text{ where e.g. } \begin{cases} P = u \\ Q = \frac{1}{2}u^2 \end{cases} \text{ or } \begin{cases} P = \frac{1}{2}u^2 \\ Q = \frac{1}{3}u^3. \end{cases} \tag{1.2a,b}$$

This conservation law itself is derived from an integral formulation of a conservation property, of mass, momentum or energy; i.e. (1.2) is the condition that

$$\int P dx - Q dt = 0 \tag{1.3}$$

around any sufficiently smooth contour in the (x,t) plane. Once we have retraced the modelling steps to (1.3), it is easy to see that one condition, called a Rankine-Hugoniot condition, which must hold across a shock wave if the conservation argument is to apply is

$$\frac{dx}{dt} = \frac{[Q]}{[P]}, \tag{1.4}$$

where [] denotes the difference between limits as the shock is approached from either side.

The result (1.4) can be derived no less reliably by the physical device of balancing fluxes in a small "control volume" surrounding the shock wave. Indeed the advantage of this device over that of retracing steps from (1.2) to (1.4) can be seen from the caution which is needed to distinguish (1.2a) from (1.2b), they both lead to the same classical partial differential equation (1.1) but correspond to quite different Rankine-Hugoniot conditions. There is a further pitfall concerning (1.4). Even though it might be thought intuitively that a single condition of this type would be sufficient to determine the single function $x = s(t)$ giving the shock position, it turns out that in allowing (1.4) we have cast our net too far in that there are an infinite number of piecewise smooth solutions of (1.1) satisfying (1.4) at points of discontinuity. We will briefly return to the question of what further constraints are necessary to ensure uniqueness in Chapter II.

2. FREE AND MOVING BOUNDARY PROBLEMS INDUCED BY A SUDDEN CHANGE IN PHYSICAL PROPERTIES

This is the class of problems with which we will be almost exclusively concerned henceforth. We cannot hope to list the enormous number of problems of this type arising in the physical and biological sciences and in the area of nonphysical applied mathematics, so we will proceed by describing a few problem areas which have been fairly intensively studied and associating with each some related problems and generalisations.

In several examples in the list below, the appearance of unknown boundaries can also be thought of as resulting from a breakdown of the type described in section I above. However, we will here be primarily considering models whose physical predictions might be unacceptable because of experimentally observed sudden changes in physical properties rather than because of the appearance of mathematical singularities. Also we will be drawing analogies between the shortcomings of models which predict liquid temperatures below the freezing point of the liquid and models which predict negative concentrations.

We must reluctantly remark that while sudden changes in physical properties pose fascinating problems at molecular levels, we will be exclusively concerned with macroscopic models in which the changes can be modelled by sharp discontinuities.

A. Frictional and Contact Problems in Solid Mechanics

Ordinary differential equations with smooth enough coefficients cannot have solutions with jump discontinuities although, for example, very jerky "relaxation" oscillations can occur in certain parameter ranges. The situation is different in the case of oscillations under the action of frictional forces, which leads to one of the relatively few ordinary differential equation models where discontinuities occur as a result of a sudden change in the physical properties of the system being modelled. Here the sudden change occurs because of the phenomenon of "limiting" friction.

In the simplest possible situation (Fig. 1) a block rests on a conveyor belt and is attached by a spring to a fixed wall. When the block is moving

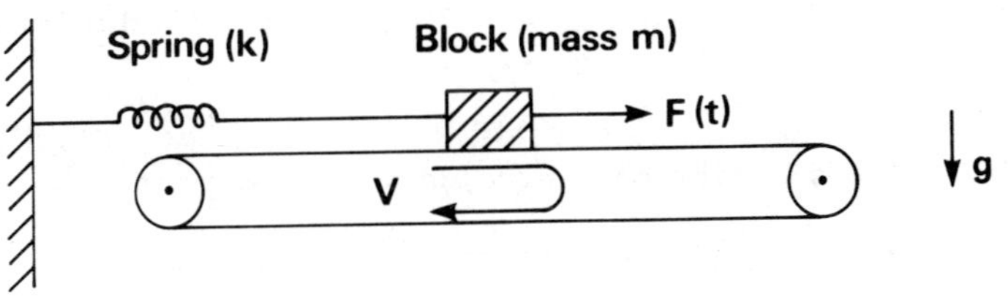

Fig. 1. Frictional Oscillator

under the action of (i) the spring, (ii) a prescribed force F(t), and (iii) "Coulomb" friction, its equation of motion is

$$m \frac{d^2x}{dt^2} + \mu mg \operatorname{sgn} \left(\frac{dx}{dt} - V\right) + kx = F(t) \tag{1.5}$$

where μ is the coefficient of friction. This equation has to be supplemented by extra information during those unknown time intervals $t_i \leq t \leq t_{i+1}$ at which $dx/dt = V$. In such an interval the block remains stationary relative to the belt until such time as $|F-kx| > \mu mg$. The set $\{t = t_i\}$ can be thought of as an unknown boundary. As in (1.4) extra information is required to determine t_i, which in this case comes from conservation of momentum: x and dx/dt are continuous functions of time at $t = t_i$, assuming that F is sufficiently smooth.

Even this very simple model exhibits many of the unpleasant features of free and moving boundary problems which we will repeatedly encounter throughout this book. Despite its simplicity when $dx/dt \neq V$, (1.5) is very tedious to solve analytically because of the difficulty of computing t_i, which may appear to be randomly distributed even when F is as simple as $\cos t$. Equally, the numerical solution of (1.5), as an initial value problem is hampered by the discontinuities in d^2x/dt^2 at $t = t_i$.

Generalisations of this problem occur in hydraulic control systems which may have many components in which there are hydraulic resistances which can be modelled as frictional resistances. Another mechanical generalisation arises in the theory of shimmy of castor wheels [191] and further examples are given in [208].

A quite different type of frictional problem which has received considerable attention recently [90, 187, 240] is that of contact between a linear elastic body in equilibrium and a rigid supporting body (Fig. 2). The displacements are assumed to be sufficiently small that all boundary conditions can be applied at the undisplaced boundary S of the elastic body, but contact may be lost over some regions S_1. On these regions the surface stress or "traction" is zero. The contact region consists of subregions S_2 over which sliding has taken place during the approach to equilibrium and S_3, where the elastic body has adhered to the support. It is the demarcations between

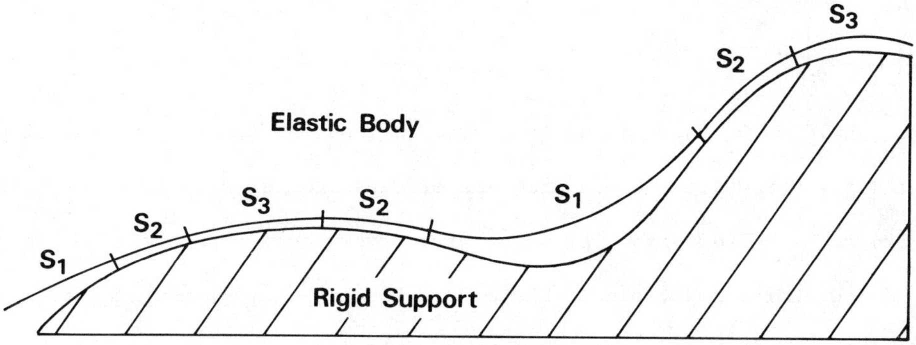

Fig. 2. Contact Problem

regions S_i which comprise the free boundary for this problem. In contrast to most of the problems in this book, the dimensionality of the free boundary is two rather than one less than that of the space in which the underlying field equations hold (but see also [104]).

Let us first consider the case of zero friction so that we do not have to consider S_3. This is an example of a so-called "Signorini problem", [106, 107] and we will only describe the simplest two-dimensional situation in the absence of gravity or other body forces. Thus we assume contact is maintained by some prescribed surface force remote from S. In the elastic body we must solve the following field and constitutive equations for the stress components σ_x, σ_y, τ and the displacements $u(x,y)$, $v(x,y)$:

$$\frac{\partial\sigma_x}{\partial x} + \frac{\partial\tau}{\partial y} = 0, \quad \frac{\partial\tau}{\partial x} + \frac{\partial\sigma_y}{\partial y} = 0 \tag{1.6}$$

$$\sigma_x = 2G\frac{\partial u}{\partial x} + \lambda\left(\frac{\partial u}{\partial x} + \frac{\partial v}{\partial y}\right), \quad \sigma_y = 2G\frac{\partial v}{\partial y} + \lambda\left(\frac{\partial u}{\partial x} + \frac{\partial v}{\partial y}\right) \tag{1.7a}$$

$$\tau = G\left(\frac{\partial u}{\partial y} + \frac{\partial v}{\partial x}\right) \tag{1.7b}$$

where G, λ are positive constants. If we assume S is smooth and denote its unit outward normal and tangent by $\underset{\sim}{n}$, $\underset{\sim}{t}$, the boundary conditions on S_1 are that the traction

$$\begin{pmatrix} \sigma_x & \tau \\ \tau & \sigma_y \end{pmatrix} \underset{\sim}{n} = \underset{\sim}{T} \quad \text{say} \tag{1.8}$$

is zero, while on S_2 there is no tangential stress and negative normal stress so that $\underset{\sim}{T}.\underset{\sim}{t} = 0$ and $\underset{\sim}{T}.\underset{\sim}{n} \leq 0$. In problems of classical elasticity theory, (1.6), (1.7) are to be solved with $\underset{\sim}{T}$ or (u,v) given everywhere on S but here, instead of $\underset{\sim}{T}.\underset{\sim}{n}$ being additionally prescribed on S_2, the rigidity of the support means that $(u,v).\underset{\sim}{n}$ is prescribed instead. Also, on S_1, $(u,v).\underset{\sim}{n}$ will satisfy a certain inequality for contact to have been lost. The classical statement of the problem is completed with suitable conditions on the remaining surface of the elastic body away from S and stress and displacement continuity conditions at the "free boundary" between S_1 and S_2.

When the effects of a positive coefficient of friction μ are included, S_2 is subdivided into adhesive regions S_3 where $(u,v).\underset{\sim}{t}$ is prescribed and, if we postulate Coulomb friction, $|\underset{\sim}{T}.\underset{\sim}{t}| < \mu|\underset{\sim}{T}.\underset{\sim}{n}|$ and regions where slipping has occurred so that $|\underset{\sim}{T}.\underset{\sim}{t}| = \mu|\underset{\sim}{T}.\underset{\sim}{n}|$. The normal displacement $(u,v).\underset{\sim}{n}$ is still prescribed everywhere in the contact region. This problem is very difficult to analyse even if the total contact area $S_2 \cup S_3$ is known in advance, as it might be if the rigid support had corners [240]. Thus, instead of discussing the effects of friction further, we will describe next the perfectly smooth elastic contact problem of a string or membrane. Although this is a very simple situation, it provides a clear motivation for the mathematical idea of a "variational inequality" which we will develop in Chapters II and IV.

Suppose that in Fig. 3, AB is a taut elastic string which is displaced from the line AB by an obstacle $z = \psi(x)$ which is smooth, both geometrically and mechanically. Then, assuming small displacements, the equation of the string is $z = u(x)$ where

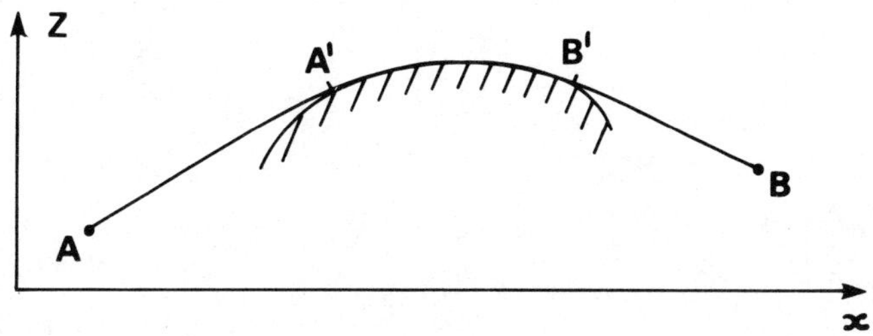

Fig. 3. Obstacle Problem for a String

$$\frac{d^2u}{dx^2} = 0 \quad \text{in AA', BB'} \tag{1.9a}$$

and

$$u = \psi \text{ in A'B'} \tag{1.9b}$$

with u and du/dx being continuous at the unknown boundary A', B'. As expected, there are more conditions at A', B' than there would be if their position was prescribed in advance. Also we note that

$$\frac{d^2u}{dx^2} \leq 0 \quad \text{in} \quad \text{A'B'}. \tag{1.9c}$$

If AA', BB' are cross sections of a membrane inflated from above by a constant pressure $-f$, the linearised membrane displacement $z = u(x,y)$ similarly satisfies, with a suitable scaling,

$$u = \psi(x,y) \quad \text{or} \quad \nabla^2 u = -f, \tag{1.10}$$

according to whether the membrane is in contact with the obstacle or not. Also u and ∇u are continuous at the unknown boundary whose cross section is A', B'.

Both (1.9) and (1.10) are special cases of the Signorini problem mentioned above but (1.10) is also reminiscent of the problem of elastic-plastic torsion of a long simply connected elastic cylinder (Fig. 4). The torsion is applied at the ends of the cylinder, whose curved sides are stress-free.

In the elastic region we need the three-dimensional version of (1.6) which we denote by

$$\frac{\partial \tau_{ij}}{\partial x_j} = 0 \qquad (1.11)$$

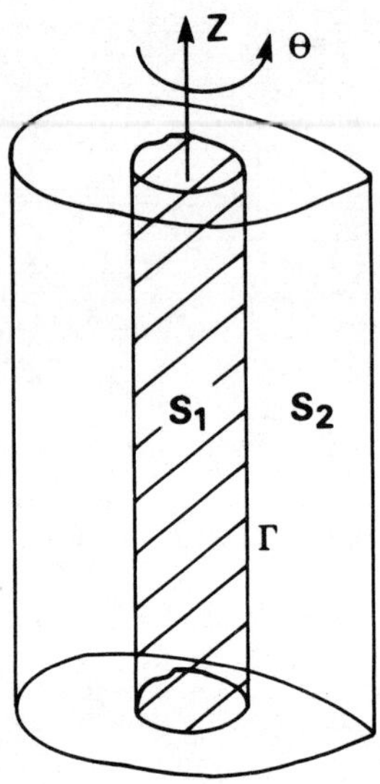

Fig. 4. Elastic/plastic Torsion Problem. The shaded region S_1 is elastic.

where τ_{ij} is the stress tensor and is symmetric. If we guess that the displacement field is $u = -\theta yz$, $v = \theta xz$, $w = w(x,y)$, which is consistent with a twist per unit length θ of the cylinder about the z-axis, we find from the three-dimensional generalisation of (1.7) that

$$\tau_{xx} = \tau_{yy} = \tau_{zz} = \tau_{xy} = 0.$$

Then (1.11) ensures the existence of a stress function $\phi(x,y)$ such that $\tau_{xz} = \partial\phi/\partial y$, $\tau_{yz} = -\partial\phi/\partial x$. A final substitution into the constitutive relations (1.7) shows that

$$\nabla^2\phi = -2G\theta \qquad (1.12)$$

in S_1. A relatively simple model for the plastic region S_2 is the Mises yield criterion

$$|\nabla\phi|^2 = k^2 = \text{const.} \tag{1.13}$$

A critical account of this criterion is given in [194] where it is also shown that the outcome of a stress balance across the unknown boundary Γ is that ϕ and $\partial\phi/\partial n$ should be continuous there. On the known stress-free outer boundary of S_2, $\tau_{xz}dy - \tau_{yz}dx = 0$ so that ϕ is a constant, which can be taken to be zero.

Since (1.13) is the equation of the surface of a pile of idealised granular material in limiting equilibrium, this torsion problem can also be interpreted as a membrane problem (1.10) in which the obstacle consists of such a granular pile [135, p. 183; 223].

We will not discuss any of the numerous other free and moving boundary problems which arise in the theory of solid mechanics here. Some of these problems, particularly those involving plates, can be treated by the ideas we will develop in subsequent chapters [107, 237], but others of great practical importance, such as models for cracks and crack propagation cannot. A review of one-dimensional models for a variety of such problems has been given in [44].

B. Problems in Heat Conduction

The Stefan problem of the melting of ice or freezing of water has been the inspiration for more research into free and moving boundary problems than most of the other problems described in this book. In its simplest form, all phenomena except that of heat conduction are neglected so that the temperature $u(x,t)$ in the ice or water phases satisfies

$$\frac{\partial}{\partial x}\left(k\frac{\partial u}{\partial x}\right) = \rho c \frac{\partial u}{\partial t} \tag{1.14}$$

away from a phase boundary in a one-dimensional situation, where the thermal conductivity k, the density ρ and the specific heat c are positive constants.

The simplest model for the phase change is to assume that it takes place at a constant temperature u = 0 with a release or uptake of a known constant latent heat L at the phase boundary x = s(t):

$$\left[k \frac{\partial u}{\partial x}\right]_{\text{solid}}^{\text{liquid}} = -\rho L \frac{ds}{dt} . \tag{1.15}$$

Since $-k\ \partial u/\partial x$ is the heat flux, (1.14) expresses conservation of energy in the body of the material and (1.15) expresses conservation of energy at the boundary. The condition u = 0 at x = s(t) can then be regarded as an equation of state. As in the mechanical problems above, an extra condition is specified at the unknown boundary over and above what would be specified at a prescribed moving boundary to compensate for the fact that s is unknown.

This Stefan problem is exceptional in that it admits one of the few useful explicit analytical solutions in the whole theory of free and moving boundary problems. This is the so-called Neumann solution and is a similarity solution when a half space of material x > 0 at constant temperature has its phase changed by the application of a different constant temperature on x = 0. Then $s = 2a\sqrt{t}$ where the constant a satisfies a transcendental equation [50]. Perhaps the most interesting feature of this solution is that a is determined uniquely in all cases except when L < 0, in which case, since changing the sign of L is the same as changing the sign of u, the material could be thought of as being superheated or supercooled. Then, if $|L|$ is sufficiently small, there is no real solution for a, and we will refer back to this situation later. In particular, we will show in Chapter V that (1.14, 1.15) is unlikely to be a realistic model for predicting the irregular phase boundaries which can occur in superheated or supercooled material.

There is a special terminology for Stefan problems of the type described above when, say, the phase boundary is moving into liquid at uniform temperature 0+ or solid at uniform temperature 0−. In these cases, the temperature in the water or ice stays at 0± respectively and u needs only to be determined on one side of the phase boundary. Such a problem is called a *one-phase* Stefan problem and more general unknown boundary problems to be solved only on one side of the unknown boundary are referred to as one-phase problems.

These "classical" Stefan problems are prototypes for many more complicated models in heat transfer. We may clearly generalise to more space dimensions, when (1.14) and (1.15) become

$$\nabla.(k\nabla u) = \rho c \frac{\partial u}{\partial t} \; ; \quad [k\nabla u.\nabla f]_{\text{solid}}^{\text{liquid}} = \rho L \frac{\partial f}{\partial t} \text{ on } f(x,y,z,t) = 0. \tag{1.14a, 1.15a}$$

We can also allow any of the coefficients to be prescribed functions of position, time or temperature, or allow the melting temperature to vary in a prescribed manner. Further, instead of driving the temperature field by the initial conditions and boundary conditions at fixed boundaries, we can introduce body heating, for example as a model for Joule heating, which gives rise to a source term in (1.14). Alternatively we could provide an extra energy source Q at the phase boundary as in laser welding problems [7] or ablation [31] problems where, for an ablating solid, the liquid phase is removed instantaneously and u has only to be found in the solid. In this one-phase Stefan problem, (1.15) is replaced by

$$\left(k \frac{\partial u}{\partial x}\right)_{\text{solid}} = \rho L \frac{ds}{dt} + Q. \tag{1.16}$$

A more drastic generalisation occurs when there is a density change at the melting temperature. Mass conservation at the phase boundary then causes flow in the liquid phase, and this is often a very important feature of practical phase change problems [210]. Liquid flow will also occur when pressure variations are important and these can also cause the phase change temperature to vary as in the theory of regelation [88, 176]. However, since a discussion of the Navier-Stokes equations is inappropriate here, we will ignore these flow problems except when they can be modelled as simply as in the theory of ablation.

C. Problems in Molecular Diffusion

Suppose first that a pure crystal is growing in a stationary liquid (the liquor) in which the concentration of the chemical of which the crystal is made is $c(x,t)$ in a one-dimensional model. Thus, in the liquor $x < s(t)$

$$\frac{\partial c}{\partial t} = \frac{\partial}{\partial x}\left(\gamma \frac{\partial c}{\partial x}\right) \tag{1.17}$$

where the diffusion coefficient γ is positive. The boundary conditions at the surface of the crystal $x = s(t)$ are more controversial than in some of the earlier models but one possibility is to assume that the liquor concentration tends to a constant equilibrium value, c_0, as $x \to s-0$. This constitutive assumption means that

$$c \to c_0 \quad \text{as } x \to s-0 \tag{1.18}$$

and since, from Fick's law, the mass flux is $-\gamma\, \partial c/\partial x$, conservation of mass at $x = s$ implies that

$$-\gamma \frac{\partial c}{\partial x} \to (c_0-1)\frac{ds}{dt} \quad \text{as } x \to s-0. \tag{1.19}$$

Equations (1.18, 1.19) may be modified to model the surface chemistry more accurately [186] but as they stand (1.17)-(1.19) comprise a one-phase Stefan problem. Since $c > c_0$ in the liquor, c can be identified as a liquid temperature but since $c_0 < 1$ the latent heat is negative and this fictitions liquid is supercooled. More realistic models for crystal growth are reviewed in [144].

Next suppose a gas such as oxygen is diffusing through a biological cell with its concentration in a one-dimensional model being $c(x,t)$. Suppose also that it is being removed in bulk from the cell at a prescribed rate R [61, 72]. This might be by some external radiation or internal chemical reaction whose resulting mass depletion effect is known a priori. Then c satisfies the diffusion equation

$$\frac{\partial c}{\partial t} = \frac{\partial}{\partial x}\left(\gamma \frac{\partial c}{\partial x}\right) - R \tag{1.20}$$

together with suitable initial and boundary conditions. Now if R is such that (1.20) would predict negative and thus physically unrealistic values for c after a certain time has elapsed, then presumably a region of very low concentration is indicated. This we model by a region of zero concentration bounded by a moving boundary $x = s(t)$, the oxygen being in $x < s(t)$. We may draw an analogy here between a model which predicts negative concentrations in the absence of a moving boundary and a Stefan problem model which predicts sub-zero water temperatures in the absence of a phase change. At $x = s(t)$, we now apply conditions similar to (1.18, 1.19). We first assume that

$$c \to 0 \text{ as } x \to s-0 \tag{1.21}$$

and then mass conservation implies that

$$\left.\frac{\partial c}{\partial x}\right|_{x=s-0} = 0 . \tag{1.22}$$

As in the case of the Stefan problem, we have two conditions at $x \to s-0$ to compensate for the fact that s is unknown.

Generalisations can be made to both these models by allowing the parameters to vary or by increasing the dimensionality of the problem. Also it is interesting to note that if we formally take the time derivative of (1.20)-(1.22), we retrieve (1.17)-(1.19) with a slight change of notation, and we will return to this transformation in subsequent chapters.

D. Problems in Incompressible Hydrodynamics

Even within this very restricted area of fluid dynamics, there are far too many free and moving boundary problems to be listed here. Explicit analytical progress is only possible for either nearly inviscid high Reynolds number flows or slow low Reynolds number flows.

a) Inviscid Flow. In the particular case of irrotational flow the field equation is Laplace's equation for the velocity potential ϕ, which is such that the velocity $\underset{\sim}{q} = \nabla\phi$. This is a possible consequence of the Euler equations of conservation of mass and momentum for an inviscid incompressible fluid. At an unknown boundary adjacent to a very low density fluid, such as the surface of the ocean or of a bubble or cavity, we impose conditions of conservation of mass and momentum. The former yields, in two-dimensional flow with the boundary being $f(x,y,t) = 0$,

$$\frac{\partial f}{\partial t} + (\underset{\sim}{q}.\nabla) f = 0. \tag{1.23}$$

The latter is equivalent, in the absence of surface tension, to applying continuity of pressure. Now the pressure p is given in terms of ϕ and the density ρ by Bernoulli's equation, which is also a consequence of the Euler equations in the body of the fluid:

$$-p = \rho gy + \rho \frac{\partial\phi}{\partial t} + \tfrac{1}{2}\rho|\nabla\phi|^2 + \text{const.} \tag{1.24}$$

Here we are accounting for gravity, with the y-axis vertically upwards, and without loss of generality the second boundary condition becomes

$$gy + \frac{\partial\phi}{\partial t} + \tfrac{1}{2}|\nabla\phi|^2 = 0. \tag{1.25}$$

The conditions (1.23) and (1.25) can be thought of as "Rankine-Hugoniot" conditions for the equations of continuity and momentum in conservation form, with the density ρ vanishing on one side of $f(x,y,t) = 0$, [204].

This combination of unknown boundary conditions for Laplace's equation yields a model which is capable of predicting a great variety of physical phenomena, ranging from river bores to tidal waves and solitons. It is therefore all the more unfortunate that except when $g = 0$ and $\partial f/\partial t = \partial\phi/\partial t = 0$ there is so little mathematical understanding of this model.

In the "two-phase" version of such problems, when there are fluids of comparable density on either side of the unknown boundary, (1.23) holds and p is continuous as the boundary is approached from either side. Thus tangential velocity discontinuities are permitted and the problem can be thought of as a "vortex sheet". The theory of the stability of such boundaries, which is part of the so-called Kelvin-Helmholtz theory, has been studied intensively, [53, 165], but again little is known about fully nonlinear solutions.

More general vortex flow may also be considered if we relax the assumption that $\underset{\sim}{q} = \nabla\phi$ everywhere away from vortex sheets, [112]. The model is easiest to formulate in two-dimensional or axisymmetric situations, for example "smoke-ring" flows, when we may introduce a scalar stream function ψ. In the two-dimensional case $\underset{\sim}{q} = (\partial\psi/\partial y, -\partial\psi/\partial x)$ where the vorticity $\omega = -\nabla^2\psi$. It is another consequence of Euler's equations that $\omega = \omega(\psi)$ so that if all the vorticity present is confined within the region $\psi \leq 0$, the free boundary problem for the streamline $\psi = 0$ is

$$\nabla^2\psi = \begin{cases} 0 & \psi > 0 \\ -\omega(\psi) & \psi < 0 \end{cases} \tag{1.26}$$

where ω is a function prescribed from the initial conditions, and which may even have a step or jump discontinuity at $\psi = 0$. In the absence of any vortex sheets, the Rankine-Hugoniot conditions of conservation of mass and momentum are that ψ and $\nabla\psi$ should be continuous across $\psi = 0$. A problem similar to (1.26) also arises in the theory of plasma confinement but the derivation of the model is too complicated to describe here [235].

Despite the lack of knowledge about harmonic functions satisfying (1.23) and (1.25), or functions satisfying (1.26) a great deal of understanding about surface waves, bubbles and vortex sheets and flows has been gained from approximate results based on probable asymptotic expansions of exact solutions [228, 248]. The appropriate asymptotic technique depends crucially on the presence of the free or moving boundary and, as stated earlier, the fruits of such an analysis often form the best basis for an intuitive understanding of the problem concerned. However, as a result of these approximations the unknown boundary is, to lowest order, either replaced by a known one or else it appears as a dependent variable in a partial differential equation. In either case, the unknown boundary aspect of the problem is drastically altered and for this reason we will not pursue these approximate theories here.

b) Slow Flow. Here the underlying equations are no longer the Euler equations but instead the Navier-Stokes equations linearised about a state of rest. In two-dimensional flow they imply the existence of a stream function ψ such that the velocity $\underset{\sim}{q} = (\partial\psi/\partial y, -\partial\psi/\partial x)$ where ψ now satisfies the biharmonic equation $\nabla^4\psi = 0$. (1.23) still applies at a boundary with a vacuum but now there are two momentum conservation conditions which result in expressions for ψ which are too lengthy to be quoted here [118]. Despite the great importance of such problems, say in glass manufacture, they are difficult to solve numerically and quite intractable analytically.

In this unsatisfactory situation we can now, in contrast to the inviscid flows mentioned earlier, profitably turn our attention to simpler approximate models for which the full biharmonic equation does not have to be solved. These approximate models will still have an intrinsic "unknown boundary" aspect to them, but they will only apply to slow viscous flows in which the fluid is confined to regions which are very thin compared to their lateral extent. Suppose, for example, the fluid is contained between two rigid walls; $y = 0$ which is stationary, and $y = h(x) > 0$ which is moving in the x direction with speed U. Then if h is small compared to the x-scale under consideration, the flow is nearly unidirectional in the x-direction. When the flow is also time independent and the appropriate asymptotic expansions are carried out on the Navier-Stokes equation in terms of h, we find from conservation of momentum that the pressure p depends only on x to lowest order and that when we have scaled the viscosity suitably, the x-velocity $u = y(y-h)/2 \; dp/dx + Uy/h$. The classical equation of this "lubrication theory" is now obtained from the integrated equation of conservation of mass, namely $d/dx \int_0^h u(x,y) \, dy = 0$. This yields

$$\frac{d}{dx}\left(\frac{h^3}{6}\frac{dp}{dx}\right) = U\frac{dh}{dx} \tag{1.27}$$

which is called Reynold's equation [45]. This equation is easy to integrate twice but in a typical situation, say in a bearing, it may predict pressures sufficiently low for cavities to form. This is analogous to the situation in B above and, without the introduction of a cavity region, say at constant pressure and with an unknown boundary, the model predicts results at variance with those observed experimentally. However, the boundary conditions at the cavity are controversial [68] and this is not the place to discuss the various models which have been proposed but we will merely state that the simplest model yields the so-called "Reynolds conditions" $p \to$ const, $dp/dx \to 0$ as the cavity is approached.

Again the problem can be generalised in various ways; apart from more sophisticated cavity analyses [214], porosity or elasticity of the walls could be incorporated [234] as well as two-dimensional effects. We will pursue the latter in the simplest case h = const., U = 0. In this case (1.27) generalises to

$$\nabla^2 p = 0 \tag{1.28}$$

which now gives the pressure distribution in fluid flow between slightly separated parallel plates. Such a configuration is called a Hele-Shaw cell. Now if there are cavities in the fluid, or if the fluid consists of blobs (Fig. 5) which are driven by internal sources or sinks, we have a

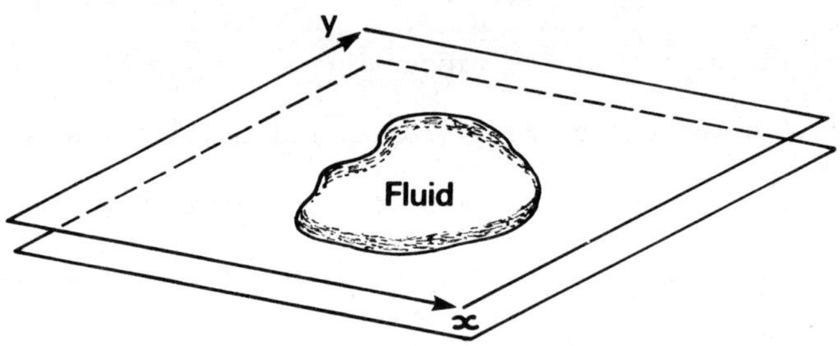

Fig. 5. Hele-Shaw Cell

boundary at which two conditions must be applied. As in the theory of lubrication cavities, three-dimensional and surface tension effects are important, making both these conditions controversial [192]. The simplest model which has been proposed, [155, 213], assumes

(i) the moving boundary has a nearly constant radius of curvature in planes perpendicular to the plates containing the mean direction of motion of the boundary. Assuming this radius is much smaller than that in planes (x,y) parallel to the plates, we may denote the projection of the moving boundary in the (x,y) plane by f(x,y,t) = 0 and a momentum balance gives

$$p = \text{const. at } f = 0. \tag{1.29}$$

This constant may be set equal to zero by trivially redefining p.

(ii) the fluid particles at f = 0 are moving with the mean fluid velocity $-h^2 \nabla p/12$. Thus, if we apply the argument leading to (1.23), we find that

$$\frac{\partial f}{\partial t} = \nabla p . \nabla f \tag{1.30}$$

after normalising so that $h^2/12 = 1$. If we write $f = p$, as we may in view of (1.29), then (1.30) becomes

$$\frac{\partial p}{\partial t} - |\nabla p|^2 = 0. \tag{1.31}$$

In justifying the simultaneous use of (1.28) and (1.31), an asymptotic analysis involving time scales is needed to confirm that time derivatives should not also appear in (1.28).

It is instructive to compare (1.29) and (1.31) with (1.23) and (1.25) with $g = 0$; the appearance of only one time derivative in the former pair of conditions makes the Hele-Shaw model much the easier to analyse, [97, 126]

Our justification for describing this relatively obscure situation in such detail is that the model dovetails with others to be described shortly in connection with porous flow. Moreover, it is important in its own right in connection with injection moulding [199] and also as a model for the theory of electrochemical machining [154]. Here an electrolyte occupies the space between a fixed cathode and an anode which is being machined or formed at a rate v_n which is experimentally observed to be proportional to the magnitude of the local current density. Assuming the process is slow enough for the electric field $\nabla\phi$ in the electrolyte to be static, then

$$\nabla^2 \phi = 0 \tag{1.32}$$

and the current $\underset{\sim}{j} \propto \nabla\phi$. In a two-dimensional situation the anode surface $f(x,y,t) = 0$, in addition to being an equipotential, is such that, by Faraday's law,

$$v_n = \frac{\partial f/\partial t}{|\nabla f|} \propto j_n \propto \frac{\partial \phi}{\partial n} \tag{1.33}$$

and we can rescale to retrieve (1.30) with a positive or negative sign depending whether the anode is contracting or expanding, i.e. whether machining or forming is taking place.

It is also of interest to note that if, in the one-phase two-dimensional Stefan problem (1.14a, 1.15a) we set the specific heat $c = 0$, or equivalently rescale t with L and let $L \to +\infty$, we retrieve the Hele-Shaw model for the temperature u. Equation (1.30) is then analogous to the Stefan condition (1.15a) as long as p is interpreted as the liquid temperature. However, only when the moving boundary is advancing, making p everywhere greater than its value at the boundary, is the liquid not supercooled. Equally when the potential-ϕ in (1.32, 1.33) is interpreted as a liquid temperature, the liquid is only not supercooled when machining rather than forming is taking place. These analogies have been discussed in [205].

E. Problems in Porous Media

The study of the flow of liquids and gases in porous media has also proved very fruitful as a source of free and moving boundary problems. Underlying much of the modelling is the well-tried D'Arcy law [171] relating the velocity $\underset{\sim}{q}$ to the pressure p in a saturated porous medium, whose pore spaces are full of fluid, with no cavities. This law states that

$$\underset{\sim}{q} = -K\nabla(p + \rho g y) \tag{1.34}$$

where y is the coordinate vertically upwards, ρ the fluid density and, in the simplest cases, K a positive constant called the permeability [22, 193]. In the case of an incompressible fluid in which $\nabla.\underset{\sim}{q} = 0$, we retrieve

$$\nabla^2 p = 0. \tag{1.35}$$

The boundary conditions at the interface $f(x,y) = 0$ between a saturated region and a completely dry region in two-dimensional steady flow are that

$$\nabla f.\nabla(p + \rho g y) = 0 \tag{1.36}$$

which is the mass conservation condition (1.23) again, and, from momentum conservation in the absence of surface tension,

$$p = 0. \tag{1.37}$$

There are clear similarities between this problem and that of Hele-Shaw flow. In particular, *unsteady* incompressible flow in a saturated porous medium is still governed by (1.35, 1.37) but the mass conversion condition (1.36) becomes

$$\frac{\partial f}{\partial t} - K\nabla(p + \rho gy).\nabla f = 0. \tag{1.38}$$

This is the same as the condition (1.30) for vertical Hele-Shaw flow when gravity is included. Moreover, if we seek a "travelling wave" solution of (1.28)-(1.30) in the form $p = p(x,\xi)$, $f = f(x,\xi)$ where $\xi = y + Ut$ and U is a constant, we find that, at $f = 0$

$$\nabla' f.\nabla'(p + U\xi) = 0, \quad \nabla' = (\frac{\partial}{\partial x}, \frac{\partial}{\partial \xi}) \tag{1.39}$$

which is (1.36) when we identify $\rho g\ \partial f/\partial y$ with $U\ \partial f/\partial\xi$. This implies that travelling wave solutions of this form in which the velocity of the moving boundary normal to itself points outwards from the flow region are analogous to porous medium flows in which the gravity field points into the saturated region.

Many generalisations again suggest themselves. For example the permeability K may vary with position or the porous medium may contain several immiscible liquids, some of which may only partially saturate the matrix [251]. Equally we may have the percolation of a compressible gas in part or whole of the region. Then even one-dimensional flow presents an interesting moving boundary problem for suppose ρ is the gas density and $p \propto \rho$ in an isothermal flow. Mass conservation requires that

$$\frac{\partial \rho}{\partial t} = \nabla.(\rho\nabla\rho) \tag{1.40}$$

with a suitable normalisation. (1.40) is a special case of the "porous medium" or percolation equation, [8]. At the boundary of the gas we assume, as in (1.37) that $\rho = 0$ but it is less clear what the condition of mass conservation is, or indeed if it is needed at all. We will return to this point in Chapter II.

As well as being a model for the effect of compressibility in porous medium flow, (1.40) has other links with some of the physical situations mentioned earlier. In particular, if an incompressible fluid occupies a shallow saturated layer of depth ρ in a porous medium above a horizontal impermeable base (Fig. 6), then the "Dupuit" or hydraulic approximation, which exploits

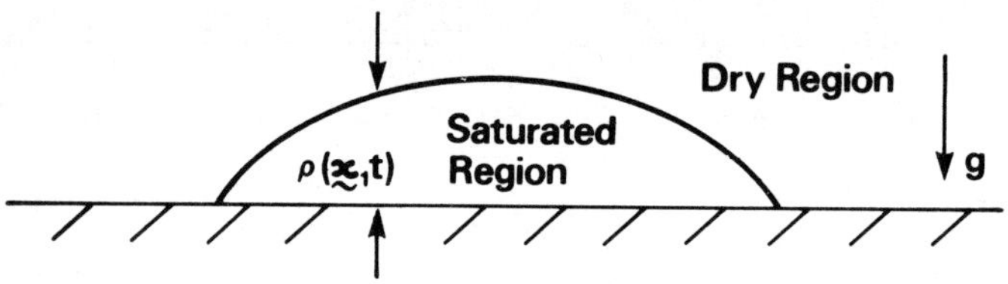

Fig. 6. Shallow Porous Medium Flow.

the shallowness of the layer may be used to show that ρ satisfies (1.40) with a suitable normalisation [218]. Equally, if a viscous liquid spreads under gravity on a horizontal base and is shallow enough for lubrication theory to be valid, its depth ρ satisfies

$$\frac{\partial \rho}{\partial t} = \nabla(\rho^3 \nabla \rho), \tag{1.41}$$

again with a suitable normalisation [42].

3. FURTHER REMARKS

In making out the above list, we must emphasise that we have only mentioned some of the archetypal free and moving boundary problems. For reasons of space and clarity we have ignored at least three sources of such problems which are stimulating much current research.

First we have made no mention of moving boundary problems which arise as limits of complicated continuous models as certain parameters tend to limiting values. For example certain types of flame can be regarded as moving boundary problems if the chemical reaction only takes place in thin layers.

This will often happen for example if the activation energy is sufficiently large. Then the theory of matched asymptotic expansions can be used [43, 160] to derive field equations away from the flame whose solutions describe perfectly smooth concentrations and temperatures. If a local analysis within the flame can be accomplished, it provides appropriate jump conditions across the flame which are of course much more complicated than the "Rankine-Hugoniot" conditions we have encountered above. In a similar vein, reaction terms in biological reaction/diffusion systems have been modelled as step functions, leading to moving boundary problems for the location of the step [201].

Secondly, we have hardly discussed any free or moving boundary problems with coupled field equations, even though almost any combinations of A-E can occur in many areas of the physical sciences. Typical examples are those of alloy solidification, which combines B and C [10, 190] and frost heaving of soils, which combines B and E [153]. As stated in the introduction, there is unfortunately no reference where the reader can find a discussion of the relative complexity of even some of these different combined areas and of which are the most suitable cases for analytical or numerical treatment.

Thirdly, we have ignored free and moving boundary problems outside the physical and biological sciences. In particular the theory of optimal stopping times [241] leads to problems similar to the Stefan problems described above with the field equation being the relevant Kolmogorov equation.

Finally, we remark that all the problems in this chapter have been written down in the form in which we feel their physical origins can be grasped most readily, without any regard for mathematical elegance and they may seem to be an ill-assorted jumble of mathematical statements. Indeed, we shall see later that some of the problems have no solutions as posed. However, the purpose of this book is to explain why some of these models are susceptible to a unifying mathematical theory which is of value both in identifying those which are well posed and in obtaining their numerical solutions.

II Intuitive Ideas about Weak and Variational Methods

In Chapter I we have described a variety of free and moving boundary problems in their crudest physical terms. These primitive models have all involved initial or boundary value problems for ordinary or partial differential equations in which the range or domain of the independent variables is to be determined as part of the solution. In just a few cases, it is possible to spot exact solutions of the problem as stated; the friction problem (1.5) and the lubrication problem (1.27) have explicit but cumbersome solutions and simple Stefan problems (1.14,1.15) and nonlinear diffusion problems (1.40,1.41) may sometimes have explicit similarity solutions [21, 50, 137]. We will also mention some less obvious closed form solutions in the course of this chapter but the inherent nonlinearity of all the models makes the existence or discovery of such explicit solutions very unlikely in practical situations.

The models of Chapter I do leave plenty of scope for manoeuvre and we will begin this chapter by describing some of the reformulations which have appeared in the literature. All these reformulations have many points of interest but we will not try to document them all in detail here. We will of course pay special attention to the two methods to which this book is addressed, but even here we will only try to convey the spirit of the methods in this chapter. More detailed statements and rigorous proofs of the use of these methods on some of the examples of Chapter I will be given in Chapters III and IV, the corresponding treatment of the remaining examples being indicated briefly in Chapter V.

There is one general approach to free and moving boundary problems which we will mention only occasionally when it impinges on the two methods in which we are principally interested. This is the "smoothing" or "regularisation" approach in which the unknown boundary is eliminated either by using smooth approximations to discontinuous data or by introducing dissipative mechanisms into the field equations or fixed boundary conditions. For example a Stefan problem with non-zero latent heat could be approximated by one with zero latent heat and smooth but rapidly varying specific heat, while a free boundary in a vortex flow could be smoothed by a small non-zero viscosity. Such smoothing can be a very helpful technique both as regards finding numerical solutions [162], proving existence results [149], and gaining insight into the physical mechanisms involved.

1. REFORMULATION OF FREE AND MOVING BOUNDARY PROBLEMS

(i) Integral Equation Methods

As with conventional initial or boundary value problems for differential equations, it is often convenient to rewrite free and moving boundary problems as integral equations at least in order to discuss existence and uniqueness of solutions. This can be done particularly easily for the Stefan problem (1.14, 1.15) since the moving boundary condition (1.15) simply states that there is a heat source of strength $\rho L\, ds/dt$ at $x = s(t)$. Hence, if $u_0(x)$ denotes the initial temperature in a material with unit conductivity, density and specific heat in $-\infty < x < \infty$,

$$u = u_0(x) * E(x,t) + L \int_0^t \frac{ds}{dt} E(x-s(\tau), t-\tau)\, d\tau \tag{2.1}$$

where $E(x,t) = 1/2\sqrt{\pi t}\, \exp(-x^2/4t)$ is the response to a unit heat source at $x = t = 0$ and $*$ denotes Fourier convolution. We may now apply the condition

$u = 0$ at $x = s$ to obtain a nonlinear Volterra integro-differential equation of the first kind for $s(t)$, and such equations of the second kind are also possible for one-phase problems [177]. Systems of such equations may be used to derive estimates needed to prove the existence, uniqueness, regularity and large time asymptotic structure of the classical solution of the Stefan problem, i.e. the solution of (1.14, 1.15), in one space dimension [209]. It is unfortunate that the method does not generalise easily to more space dimensions and indeed even the question of existence of the classical solution of (1.14a, 1.15a) in two space dimensions is not resolved.

Another more unusual integral equation reformulation may be derived for the one-phase Stefan problem, despite its nonlinearity, by taking a Laplace transform in time. Suppose for simplicity that we consider the one-phase problem of freezing water at temperature 0+ in $x > 0$ by applying a heat flux $g(t)$ at $x = 0$, $t > 0$ so that

$$\frac{\partial^2 u}{\partial x^2} = \frac{\partial u}{\partial t}, \qquad 0 < x < s(t), \tag{2.2}$$

$$u \to 0, \quad \frac{\partial u}{\partial x} \to L\frac{ds}{dt} \quad \text{as} \quad x \to s-0, \tag{2.3}$$

with

$$\frac{\partial u}{\partial x} = g(t) > 0 \quad \text{at} \quad x = 0,\ t > 0 \tag{2.4}$$

and

$$s(0) = 0. \tag{2.5}$$

If we assume the inverse function $\ell(x) = s^{-1}(x)$ exists and define $u \equiv 0$ in $x > s$, and put

$$\tilde{u}(x,p) = \int_{\ell(x)}^{\infty} u(x,t)e^{-pt}dt \tag{2.6}$$

then

$$\frac{d^2\tilde{u}}{dx^2} - p\tilde{u} = -Le^{-p\ell(x)} \tag{2.7}$$

with

$$\frac{d\tilde{u}}{dx} = \tilde{g}(p) \quad \text{at} \quad x = 0. \tag{2.8}$$

Then if we also make some assumptions about the growth of $\ell(x)$ as $x \to \infty$, [125, 177], (2.7 , 2.8) only has a solution if

$$\tilde{g}(p) = L \int_0^\infty e^{-p\ell(\xi)} \cosh(p^{\frac{1}{2}}\xi)\, d\xi. \tag{2.9}$$

This Fredholm integral equation for ℓ is probably only useful in its own right for deriving asymptotic estimates, but it is interesting to note that it is the Laplace transform of the infinite order ordinary differential equation

$$g(t) = L \sum_{n=1}^{\infty} \frac{1}{(2n-1)!} \frac{\partial^n}{\partial t^n} [s(t)]^{2n-1} \tag{2.10}$$

which is itself the result of applying the boundary condition (2.4) to the representation

$$u = -L \sum_{n=1}^{\infty} \frac{1}{(2n)!} \frac{\partial^n}{\partial t^n} [x-s(t)]^{2n}$$

of the solution of (2.2 , 2.3), [46].

Integral equations have proved exceptionally valuable in the theory of surface gravity waves, [54, 142] and in particular the problem of spatially periodic waves satisfying (1.23, 1.25) can be reduced to a generalised non-linear Fredholm equation. Also, frictional and contact problems can sometimes be written as integral equations or dual integral equations [224, 225]. However, we will not consider explicit integral equations further, although variational and conservation integrals will appear frequently later in the chapter.

(ii) Mapping Methods

The principal difficulty in treating any one of the problems of Chapter I would be removed if a transformation of the variables could be found which resulted in a differential equation to be solved on a prescribed domain. One of the best-known techniques of this type is for steady inviscid flows with a free surface in the absence of gravity, i.e. (1.23, 1.25) with $g = \partial f/\partial t = \partial\phi/\partial t = 0$. In this case, the stream function ψ as defined before (1.26) is the harmonic conjugate of the velocity potential ϕ and so

$$\nabla^2\psi = 0 \tag{2.11}$$

with

$$\psi = \text{const}, \quad \frac{\partial\psi}{\partial n} = \text{const}. \tag{2.12}$$

on the free surface. Also ψ is constant on any fixed boundaries which are present. The crucial observation is that if we write

$$\log\{\frac{d}{dz}(\phi + i\psi)\} = \log\frac{dw}{dz} = Q(z) \tag{2.13}$$

where $z = x + iy$, then

$$R\ell\ Q = \text{const} \tag{2.14}$$

on the free surface. Hence, if any fixed boundaries which are present are straight, so that the flow inclination and hence $\mathrm{Im}\ Q(z)$ are prescribed constants on them, then the boundaries of the flow region in the complex Q plane are known straight lines. Thus, since ψ is a harmonic function of (x,y) and hence of $(R\ell\ Q, \mathrm{Im}\ Q)$, the problem has been reduced to a Dirichlet potential problem on a prescribed domain, from the solution of which we can find $\phi + i\psi$ as an analytic function of Q. Integration of the ordinary differential equation (2.13) then yields $\phi(x,y)$ but it is usually more economical to use the Riemann mapping theorem to relate the Q and w planes to solve these "Helmholtz" flows [28].

A very similar technique is available for steady flows in porous media in the *presence* of gravity, again as long as any impermeable boundaries are straight. In contrast to (2.12) we now have, from (1.36, 1.37),

$$\nabla^2 p = 0 \tag{2.15}$$

with

$$p = 0 \quad \text{and} \quad \frac{\partial p/\partial y}{|\nabla p|^2} = \text{const.} \tag{2.16}$$

at the free boundary between the saturated and dry regions. If we again write $w(z) = p + iq$, where q is the harmonic conjugate of p, the crucial observation is now that $\text{Im}(dw/dz)^{-1}$ = const on the free boundary. Also $\text{Im}\{\log(\overline{dw/dz} - i\rho g)\}$ is the flow inclination and we have a problem with prescribed circular and straight boundaries in the dw/dz plane. The use of conformal mapping to obtain dw/dz as a function of w has been described in many works on porous medium flows [22, 193, 252]. Precisely the same technique applies to "travelling wave" solutions in Hele-Shaw cells and the equivalence of (1.36) and (1.39) has been exploited in [213]. Corresponding situations have also been treated in electrochemistry [129, 172] as have Stefan problems with zero specific heat [221].

As soon as the genuinely unsteady problem (1.29, 1.31) is considered, complex variable methods become much harder to apply. The unsteady problem of suction from or injection into an initially prescribed blob of liquid from a point sink or source inside the initial blob has been considered in [199]. Mapping the flow region in the z-plane into the unit circle in the ζ plane through an analytic function $z = f(\zeta,t)$ only reduces the problem to a functional equation of the form $g(f(\zeta)) = \overline{f(1/\bar{\zeta})}$ where the bar denotes the complex conjugate and g is an analytic function whose singularities external to the liquid region only are known. Explicit solutions can only be found in a few simple cases. Equally for unsteady irrotational free surface flows without gravity, the Wagner transformation [246] can be employed on the so-called "wedge-entry" problem (Fig. 7). This concerns the penetration of a

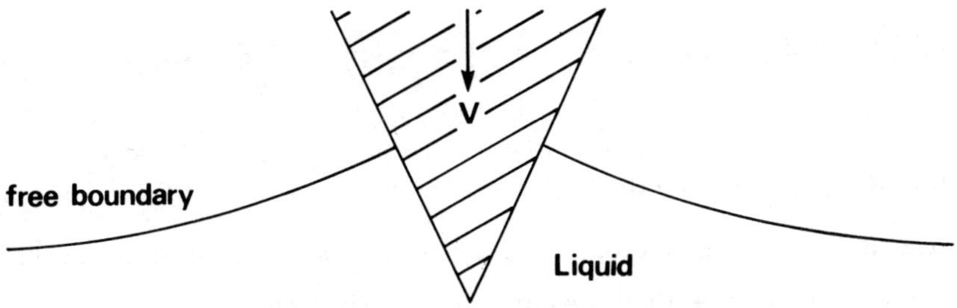

Fig. 7. Wedge-entry Problem

liquid half space by an infinite solid wedge so that there is no length scale associated with the problem. It can be reduced to a complicated integral equation for a function of a single variable [85].

Conformal mapping methods can only apply to two-dimensional free and moving boundary problems and then only when the field equation is Laplacian. For problems such as the one-dimensional Stefan problem where the phase boundary is $x = s(t)$, we may map the physical plane into one in which $x/s(t)$ is a new independent variable, which is the so-called Landau transformation [143], or into one in which u is a new independent variable, which is the isotherm-migration method [73]. The field equation in the transformed variables then either involves the phase boundary in its coefficients, or is nonlinear. Nonetheless both methods can be very effective computationally [117, 238].

The isotherm migration method is reminiscent of hodograph methods in which the components of the gradient of the dependent variable are used as new independent variables. The conformal mapping methods described earlier use such independent variables but the hodograph method coupled with the

Legendre transformation can be used to analyse much more general free boundary problems [140]. It is best known in the theory of two-dimensional, inviscid compressible flow past aerofoils. There the nonlinear equation

$$\{1 - (\frac{1}{a}\frac{\partial\phi}{\partial x})^2\}\frac{\partial^2\phi}{\partial x^2} - \frac{2}{a^2}\frac{\partial\phi}{\partial x}\frac{\partial\phi}{\partial y}\frac{\partial^2\phi}{\partial x\partial y} + \{1 - (\frac{1}{a}\frac{\partial\phi}{\partial y})^2\}\frac{\partial^2\phi}{\partial y^2} = 0, \tag{2.17}$$

where $a(\partial\phi/\partial x, \partial\phi/\partial y)$ is prescribed, has to be solved in the region exterior to the prescribed aerofoil position $y = \alpha(x)$ on which the normal derivative $\partial\phi/\partial n = 0$. If we now define the Legendre transformation $\Phi(u,v) = x\,\partial\phi/\partial x + y\,\partial\phi/\partial y - \phi$ where $u = \partial\phi/\partial x$, $v = \partial\phi/\partial y$, then (2.17) becomes the linear equation

$$(a^2 - u^2)\frac{\partial^2\Phi}{\partial v^2} + 2uv\frac{\partial^2\Phi}{\partial u\partial v} + (a^2 - v^2)\frac{\partial^2\Phi}{\partial u^2} = 0. \tag{2.18}$$

However the boundary condition $v = u\,\alpha'(x)$ on $y = \alpha(x)$ now becomes an unknown boundary condition

$$\frac{\partial\Phi}{\partial v} = \alpha(\frac{\partial\Phi}{\partial u}) \text{ on } v = u\alpha'(\frac{\partial\Phi}{\partial u}) . \tag{2.19}$$

Curiously enough, this last problem is amenable to the variational inequality technique for free boundary problems, to be described shortly [39].

(iii) Discretisations

There are too many different ways of expanding the solution of the governing primitive differential equation and the function describing the unknown boundary in terms of finite or countable sets of elementary functions for us to begin to list them here. However, several reviews of the consequent numerical advantages and disadvantages are available [69, 81, 111, 117, 164]. As stated in the introduction, these discretisations are the most commonly used and practically valuable methods for solving free and moving boundary problems.

In the case of parabolic governing equations, time-stepping methods have also been used for existence and uniqueness proofs [46, 103, 130].

(iv) Embedding and Fixed Domain Methods

We will conclude this section with a brief description of two methods, at least one of which is related in spirit to the two methods which will form the basis of this book. The idea is to remove the unknown boundary aspect not by a mapping method but by considering the problem as a member of a wider class of problems posed on a prescribed domain which entirely contains the unknown boundary. In the case of "one-phase" problems, this may involve considering problems defined on a domain consisting of both the unknown physical domain and an adjacent new domain which may have no direct physical interpretation. Instead of trying to adjust the physical domain so that the requisite conditions are satisfied at the unknown boundary, which is the approach most readily suggested by the formulations of Chapter I, the philosophy is now to adjust some property of a class of solutions to a fixed domain problem until a solution and an interior boundary are selected for which the solution happens to have the correct behaviour at that boundary.

This idea has been used in [30] to derive integro-differential equations for a variety of Stefan problems. In the simplest one-phase case, the physical domain $0 < x < s(t)$ is embedded in a fictitious thermal material in $0 < x < a$, where $s(t) < a$ for all times of interest, and a boundary condition is applied at $x = a$ involving one arbitrary function. Using the idea in (2.1), coupled equations are then derived for this function and $s(t)$.

An alternative approach has been adopted in [25, 203] to deal with the oxygen-consumption problem described in Chapter I. Instead of (1.20)-(1.22), we consider the problem

$$\frac{\partial c}{\partial t} = \frac{\partial}{\partial x}\left(\gamma \frac{\partial c}{\partial x}\right) - F(c) \tag{2.20}$$

where

$$F(c) = \begin{cases} R, & c > 0 \\ 0, & c = 0 \end{cases} \tag{2.21}$$

where x is now in the extended fixed domain $0 < x < a$, $a > s(t)$, and we also put

$$c = 0 \text{ or } \frac{\partial c}{\partial x} = 0 \text{ at } x = a. \tag{2.22}$$

This model has the reasonable physical interpretation that the consumption is switched off as soon as the concentration falls to zero and it clearly suggests a numerical discretisation in which no explicit mention is made of the position at which $c \to 0+$. This discretisation is precisely that for $\partial c/\partial t = \partial/\partial x(\gamma\, \partial c/\partial x) - R$ with the proviso that any values of c which are computed to be negative at any particular time step are automatically set equal to zero before embarking on the next time step. This discretisation is called the "truncation method" and can be justified rigorously by considering the limit of a sequence of smooth approximations to $F(c)$ such as the Michaelis-Menten consumption rate $F(c) = Rc/c+\varepsilon$, $\varepsilon \ll 1$ [203].

The truncation method involves guessing how to introduce the function F in (2.21), although there are physical clues available. There are possibly many ways in which we could similarly set about regarding the other problems of Chapter I as fixed domain problems, and what we shall now do is describe in more detail two such formulations. The first is a fixed domain formulation based on the idea of widening the definition of solutions of the field equation to allow automatically for solutions with prescribed discontinuities at the unknown boundary; the second is based initially on a variational formulation in which the class of functions over which the minimisation is to be carried out is suitably restricted.

2. WEAK FORMULATIONS OF FREE AND MOVING BOUNDARY PROBLEMS

(i) Problems of Type 1 (p. 4)

We will begin by recalling a well-known fixed domain approach to the theory of conservation laws of the form (1.2). It is inspired by the integral reformulation (1.3), but instead of regarding (1.3) as a result of integrating (1.2) over an arbitrary region of the (x,t) plane, and applying Stokes theorem, we multiply (1.2) by an arbitrary test function $\phi(x,t)$ which decays sufficiently rapidly as $x^2 + t^2 \to \infty$ for us to be able to integrate the resulting identity over the half-space $t > 0$. If ϕ is also smooth enough for us to be able to integrate by parts, we obtain

$$\int_0^\infty \int_{-\infty}^\infty \phi\left(\frac{\partial P}{\partial t} + \frac{\partial Q}{\partial x}\right) dxdt = 0 =$$

$$= -\int_0^\infty \int_{-\infty}^\infty \left(P\frac{\partial \phi}{\partial t} + Q\frac{\partial \phi}{\partial x}\right) dxdt - \int_{-\infty}^\infty P(x,0)\,\phi(x,0)\,dx. \tag{2.23}$$

This leads to the definition of a weak solution of (1.2) as a pair of functions (P,Q) such that the right-hand side of (2.23) vanishes identically for all suitable ϕ. A precise statement is given in [145].

It is a simple matter to prove that any classical solution of (1.1), by which we mean a pair of functions (P,Q) which satisfy (1.2) everywhere is necessarily a weak solution, but the converse need not be true. Moreover, if (P,Q) is a classical solution except at a curve $x = s(t)$ across which there is a step or jump discontinuity satisfying (1.4), then the right-hand side of (2.23) still vanishes identically even though (1.1) and (1.2) are not satisfied at $x = s(t)$. This is the principal advantage to be gained from the idea of a weak solution: it automatically caters for the presence of any finite number of shock waves which satisfy the Rankine-Hugoniot condition (1.4). Moreover, the weak solutions corresponding to (P,Q) defined by (1.2a) are quite different from those corresponding to (1.2b), although (1.2a, 1.2b) lead to identical classical solutions.

There is, however, one major drawback to the definition (2.23). It does not in general yield a unique weak solution. In generalising our idea of classical solutions, we have cast our net too far and, taking inviscid gas dynamics as an example [67], it can be shown that the generalisation of (2.23) allows not only the compressive shock waves which are observed physically but also expansion shock waves across which the entropy of a fluid particle would decrease. Hence, for hyperbolic equations of this type, further restrictions are needed to ensure the uniqueness of the weak solution. These restrictions may come from physical arguments involving stability, entropy or dissipation [132, 182] or continuous dependence on the initial data [146]. For example, it can be shown that solutions of the initial value problem for Burgers equation

$$\frac{\partial u}{\partial t} + \frac{\partial}{\partial x}\left(\tfrac{1}{2}\, u^2\right) = \mu\, \frac{\partial^2 u}{\partial x^2} \tag{2.24}$$

never develop singularities when the "viscosity" $\mu > 0$, but that as $\mu \to 0+$, they tend to members of a subset of all weak solutions of (1.1) within which solutions of the initial value problem are unique.

In the context of this book, the idea that there may exist finite difference approximations of (1.1) whose solutions automatically tend to the restricted unique weak solution is particularly interesting. If (1.1) is discretised so that some kind of "artificial viscosity" is introduced, it can be shown [147, 200] that the numerical solution will converge in some mean sense to the above mentioned unique weak solution. No explicit mention of shock waves is needed, but they emerge automatically in the numerical output as regions in which u changes rapidly.

To make these arguments rigorous would take us too far from our main theme of free and moving boundary problems of type II in Chapter I. Nonetheless it was the analogy with the ideas developed for hyperbolic equations that led to suggestions of a similar approach for Stefan problems [207], and the first theory was presented in [136], following [183]. A recent discussion is given in [204]. This chronological development is surprising since many of the problems of type II are easier to handle than shock waves in hyperbolic systems, simply because the uniqueness question is often easier to answer. For simple Stefan problems, no further restrictions are required to complement the integral statement analogous to (2.23) to render the weak solution unique.

(ii) Problems of Type 2 (p. 6)

(a) The Porous Medium Equation (1.40, 1.41)

For simplicity let us consider (1.40) in only one space dimension $-\infty < x < \infty$. This equation is already in conservation form so we may proceed as in (2.23) to define a weak solution as a positive continuous function $\rho(x,t)$ such that

$$\int_0^T \int_{-\infty}^{\infty} (\rho \frac{\partial \phi}{\partial t} + \tfrac{1}{2}\rho^2 \frac{\partial^2 \phi}{\partial x^2})\, dxdt = - \int_{-\infty}^{\infty} \phi(x,0)\rho(x,0)\,dx \qquad (2.25)$$

for all test functions ϕ which are sufficiently smooth and vanish as $|x| \to \infty$ and at $t = T$, [184]. The definition is restricted to finite time intervals $(0,T)$ because of possible "blow-up" of the solution in finite time, but of even more importance is the fact that the integration is over the whole x-axis including regions of the porous medium where no gas is present so that $\rho \equiv 0$.

The uniqueness of this weak solution can be proved by subtraction of (2.25) for two candidates ρ_1, ρ_2 with the same initial data and choosing ϕ to be a function of ρ_1 and ρ_2 which makes the integrand a perfect square. The proof of existence is more complicated and requires the study of the convergence of approximating sequences, [184].

Near a moving boundary $x = s(t)$ which is such that $\rho \equiv 0$ in $x > s(t)$, a careful estimate can be made of the left-hand side of (2.25) to show that, [141],

$$\lim_{x \to s-0} \frac{\partial \rho}{\partial x} = -\frac{ds}{dt} . \tag{2.26}$$

This is the Rankine-Hugoniot condition for a weak solution and its physical interpretation is that of conservation of mass. Now the argument leading to (1.19) only gives $\lim\limits_{x \to s-0} \rho\, \partial\rho/\partial x = 0$ and to derive (2.26) we must consider small quantities of second order. If we assume that $\lim\limits_{x \to s-0} \partial\rho/\partial x$ is finite and that the boundary moves from $x = s$ to $x = s + \Delta s$ in time Δt, then mass conservation gives

$$\tfrac{1}{2}\left(\left.\frac{\partial\rho}{\partial x}\right|_{x=s} \Delta s\right) . \Delta s \simeq -\int_0^{\Delta t} \rho \left.\frac{\partial \rho}{\partial x}\right|_{x=s} dt = -\left(\left.\frac{\partial\rho}{\partial x}\right|_{x=s}\right)^2 . \frac{\Delta s}{\Delta t} . \frac{(\Delta t)^2}{2}$$

so that

$$-\left.\frac{\partial\rho}{\partial x}\right|_{x=s} = \lim_{\Delta t \to 0} \frac{\Delta s}{\Delta t} = \frac{ds}{dt} .$$

This is the moving boundary condition which is always satisfied by weak solutions of (1.40) at the boundary of regions in which $\rho \equiv 0$. It is important because a classical solution of the initial value problem for (1.40) will only be the unique weak solution if it satisfies (2.26). There are many functions which satisfy (1.40) for $\rho > 0$ in $x < s(t)$ and satisfy the same initial condition but do not satisfy (2.26), [178].

For practical purposes the most important property of the weak solution is that several numerical discretisations of (1.40), if solved on a sufficiently large fixed domain $-L < x < L$, $0 < t < T$, have solutions which tend to the weak solution as the mesh size tends to zero. The most obvious scheme is to define $\frac{1}{2}\rho^2 = p$, $x = i\delta x$, $t = n\delta t$ and solve

$$(\rho_i^{n+1} - \rho_i^n)/\delta t = (p_{i+1}^n - 2p_i^n + p_{i-1}^n)/\delta x^2 \tag{2.27a}$$

$$p_i^{n+1} = \tfrac{1}{2}(\rho_i^{n+1})^2 \tag{2.27b}$$

with suitable initial conditions and conditions at the fixed boundaries $x = \pm L$. It can be shown, [197], that $\rho_i^n \to \rho(i\delta x, n\delta t)$ in some mean sense even when the exact solution has a moving boundary at which (2.26) holds, although of course (1.40) does not. Thus we can use (2.27) without any distinction between regions where $\rho > 0$ and where $\rho \equiv 0$ to calculate solutions of (1.40) in which part of the medium is unoccupied by gas. This is a useful computational result and similar schemes have even been used to describe numerically situations in which $s(t)$ has a singularity [124], but why (2.27) works at all is because of the existence and uniqueness of the weak solution (2.25).

(b) Stefan Problems

When we come to consider the simplest two-phase Stefan problem in a similar way, we could note that (1.14) appears to be in conservation form. Then, were we to use the above ideas, we would derive a Rankine-Hugoniot condition $[k\ \partial u/\partial x]_{\text{solid}}^{\text{liquid}} = -[\rho cu]_{\text{solid}}^{\text{liquid}}\ ds/dt$, which contradicts (1.15) since u is continuous at the phase boundary. This suggests that before defining a weak solution, we should rewrite (1.14) in terms of the *enthalpy* H defined by (Fig. 8a).

$$H = \begin{cases} \rho cu, & u < 0 \\ \rho cu + \rho L, & u > 0 \end{cases} \tag{2.28}$$

$$\frac{\partial H}{\partial t} = \frac{\partial}{\partial x}\left(k\,\frac{\partial u}{\partial x}\right) \tag{2.29}$$

and indeed this is the correct thermodynamic conservation law for heat con-

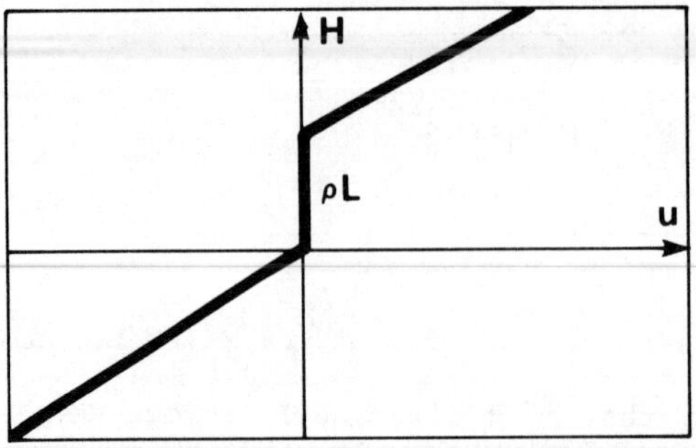

Fig. 8a

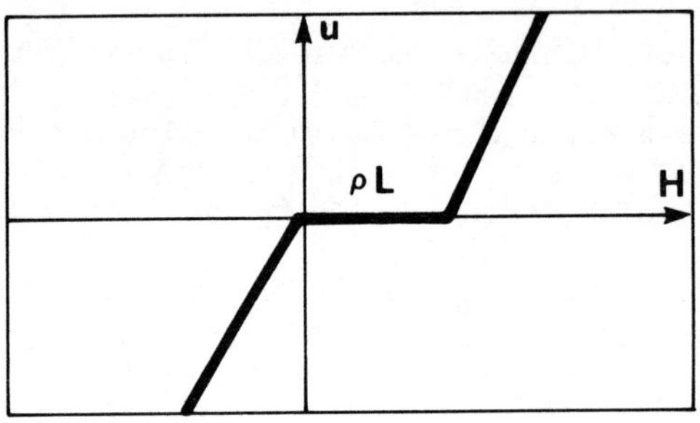

Fig. 8b

duction. (2.28) implies that the enthalpy at zero temperature can take any value between 0 and ρL and thus the temperature distribution does not define the enthalpy uniquely. However H is a monotone increasing function of u and at u = 0 it has a positive jump. Thus H defines u uniquely (Fig. 8b).

Restricting ourselves for simplicity to problems in $-\infty < x < \infty$, we now define a weak solution of (2.29) to be a pair of functions (u,H) related by (2.28) such that

$$\int_0^T \int_{-\infty}^{\infty} \left(ku \frac{\partial^2 \phi}{\partial x^2} + H \frac{\partial \phi}{\partial t}\right) dxdt = -\int_{-\infty}^{\infty} H(x,0)\,\phi(x,0)\,dx, \tag{2.30}$$

for all test functions ϕ which have continuous first time and second space derivatives and vanish as $|x| \to \infty$ and at $t = T$ [136]. By integrating by parts, we may write the first term in the integrand of (2.30) as $-k\partial u/\partial x\ \partial \phi/\partial x$ but when we use this latter expression we must interpret it correctly at $x = s(t)$. Then we can repeat the argument following (2.23) to show that the Rankine-Hugoniot condition across a curve $x = s(t)$ at which $u = 0$ becomes

$$[k \frac{\partial u}{\partial x}]_{\text{solid}}^{\text{liquid}} = - [H]_{\text{solid}}^{\text{liquid}} \frac{ds}{dt} = -\rho L \frac{ds}{dt} \tag{2.31}$$

in agreement with (1.15).

We also note that we can easily modify the above ideas to allow for more general problems in which ρ, c and k are still positive, but may depend on u. We simply let

$$H = \begin{cases} \int_0^u \rho c \, du, & u < 0 \\ \int_0^u \rho c \, du + \rho L, & u > 0 \end{cases} \tag{2.28a}$$

and make the "Kirchoff" transformation

$$\bar{u}(u) = \int_0^u k \, du . \tag{2.32}$$

Then (2.29) becomes

$$\frac{\partial H}{\partial t} = \frac{\partial^2 \bar{u}}{\partial x^2} \tag{2.33}$$

and ku is replaced by $\bar{u}$ in (2.30). The function $H(\bar{u})$ still has a jump of ρL at $\bar{u} = 0$ so that (2.31) still holds.

The uniqueness theorem for weak solutions depends crucially upon H being a monotonic increasing function of u, in the simplest case when all the thermal parameters are constant. We have already (p. 14) questioned the value of the model (1.14, 1.15) when $H(u)$ is not a monotone function and some further discussion of this situation will be given in Chapter V. However, when $L > 0$, we may subtract two solutions (u_i, H_i), $i = 1,2$, to obtain, in the case $k = 1$,

$$\int_0^T \int_{-\infty}^{\infty} \left[\frac{u_1-u_2}{H_1-H_2}\frac{\partial^2\phi}{\partial x^2} + \frac{\partial\phi}{\partial t}\right](H_1-H_2)\ dxdt = 0. \tag{2.34}$$

To prove $H_1 = H_2$ almost everywhere is now easy if we can show that any arbitrary smooth bounded function can be written in the form of the square brackets in (2.34) for some suitable ϕ. Since ϕ vanishes at $t = T$, this demonstration would involve solving a backward heat equation if $H(u)$ was not monotonic. We may also note that neither this method nor the one mentioned after (2.25) could be used to prove uniqueness for (2.23). For example, the term in square brackets in (2.34) would become $\partial\phi/\partial t - \frac{1}{2}(u_1+u_2)\partial\phi/\partial x$ and, from the theory of first order partial differential equations, it is easy to see that arbitrary smooth functions cannot be written in this form.

The existence theory for weak solutions is similar to that for (a) above, but again the most important practical result is that the discretisation suggested by (2.29), namely

$$(H_i^{n+1}-H_i^n)/\delta t = k(u_{i+1}^n-2u_i^n+u_{i-1}^n)/\delta x^2 \tag{2.35a}$$

$$u_i^{n+1} = U(H_i^{n+1}), \qquad x = i\delta x, \qquad t = n\delta t,$$

where U is defined by

$$U(H) = \begin{cases} H - L, & H > L \\ 0\ , & L > H > 0 \\ H\ , & 0 > H \end{cases} \tag{2.36}$$

converges in mean square to the weak solution, [136]. Again there is no explicit mention of the moving boundary except in the definition of $U(H)$. Although (2.35, 2.36) is an explicit discretisation, and is adequate for this problem, the implicit scheme in which (2.35a) is replaced by

$$(H_i^{n+1} - H_i^n)/\delta t = k(u_{i+1}^{n+1} - 2u_i^{n+1} + u_{i-1}^{n+1})/\delta x^2 \tag{2.35c}$$

has similar convergence properties, [58], and, as we shall see, it applies to some other types of unknown boundary problems which cannot be handled by (2.35a).

There is one situation where the fixed domain nature of this "enthalpy method" has very interesting results and implications for the model (1.14). This is the case when melting of a solid occurs not as a result of heat being conducted inwards from a fixed boundary, but rather by volumetric heating [9, 202]. An example which we will describe in more detail in Chapter V is Joule heating of a metal by the passage of an electric current, where it appears that the weak formulation may have a physical as well as a mathematical advantage over the classical one.

(c) Other Problems in Heat and Mass Transfer

The concept of the weak solution for the simplest Stefan problems is difficult to generalise to problems in which the melting temperature or latent heat are non-constant [233], or those in which other phenomena occur as well as heat conduction [77]. Also, the fact that the enthalpy H must be monotonic increasing is important when deciding which of the problems of Chapter I which seem to be related to Stefan problems can be treated by the enthalpy method. For example the Hele-Shaw problem (1.28)-(1.31) for an expanding liquid region is a one-phase Stefan problem in which the liquid temperature is p, the specific heat c is zero and $\rho L = 1$. Now the validity of some of the earlier statements about weak solutions is not clear when $c = 0$ and this will be discussed further in Chapter III. Nonetheless, numerical experiments can be performed using the implicit scheme (2.35b, 2.35c) to time step in H and the results compare well with analytical and other numerical predictions [74]. Thus we can plausibly compute the shape of expanding liquid regions in Hele-Shaw flow, but not contracting ones. The same remarks apply to electrochemical machining and forming respectively.

The oxygen-consumption problem (1.20)-(1.22) has already been mentioned in connection with the fixed domain truncation method. If we try to apply the enthalpy method by writing $u = \partial c/\partial t$, when $\gamma = 1$ we obtain

$$\frac{\partial^2 u}{\partial x^2} = \frac{\partial u}{\partial t} \tag{2.36}$$

with

$$u = 0, \quad \frac{\partial u}{\partial x} = -R\frac{ds}{dt} \quad \text{at} \quad x = s-0 \tag{2.37}$$

together with initial and boundary conditions which may involve distributions if the initial and boundary conditions for c are not smooth enough. Thus u is the liquid temperature in a one-phase Stefan problem with latent heat R. However $u < 0$ in regions of oxygen consumption so this is a supercooled Stefan problem and therefore not amenable to the enthalpy method.

(d) Frictional Oscillations

The discretisation of the enthalpy formulation for Stefan problems can be viewed in another way if we regard the field equation not as a conservation law but as a differential equation in time with discontinuous coefficients. In (1.5), the discontinuous term is undifferentiated, so to solve by a discretisation analogous to (2.35) we write $y = dx/dt - V$ so that

$$\frac{dy}{dt} + \alpha \operatorname{sgn} y + x = F(t) \tag{2.38a}$$

$$\frac{dx}{dt} - y = V. \tag{2.38b}$$

Now the function

$$\tilde{H}(y) = y + \alpha\delta t \operatorname{sgn} y + \delta t^2 y \tag{2.39}$$

is monotonic increasing when $\alpha > 0$ so that if we put $t = n\delta t$, the implicit discretisation

$$(y^{n+1}-y^{n})/\delta t + \alpha \text{ sgn } y^{n+1} + x^{n+1} = F(t^{n+1}) \tag{2.40a}$$

$$(x^{n+1}-x^{n})/\delta t - y^{n+1} = V \tag{2.40b}$$

is well defined; y^{n+1} and hence x^{n+1} can be found at each step from

$$\tilde{H}(y^{n+1}) = y^{n} - \delta t(x^{n} + V\delta t) + \delta t \, F(t^{n+1}) \, . \tag{2.41}$$

The convergence of (2.40) as $\delta t \to 0$ is proved in [95, 99], and we can regard the limit function as one possible generalised solution of (1.5). Another possibility for an initial value problem is to multiply (1.5) by a sufficiently smooth test function, vanishing together with its derivative at some time $t = T$, and integrate over $0 < t < T$ to arrive at a definition similar to (2.30). The Rankine-Hugoniot conditions emerge as the physical conditions that x and dx/dt are continuous at the "sticking" times $t = t_i$, [178]. There are also several other ways of defining generalised solutions of arbitrary ordinary differential equations with discontinuous nonlinearities but it is unclear which definition suggests the best computational method in any particular case [35, 108, 127, 230, 232].

3. VARIATIONAL METHODS FOR FREE AND MOVING BOUNDARY PROBLEMS

We have so far considered some of those free and moving boundary problems from Chapter I which can be formulated as conservation laws, thus giving a physical motivation for their study using the theory of weak solutions. We will now consider those for which some kind of variational principle may be appropriate. The most obvious candidate is the vortex flow problem (1.26).

(a) Vortex Flows and Plasma Confinement

Since the Euler equation for stationary values of the functional

$$\iint \{\tfrac{1}{2}|\nabla\psi|^2 - \Omega(\psi)\} \, dxdy \tag{2.42a}$$

where

$$\frac{d\Omega}{d\psi} = \begin{cases} 0 & \psi > 0 \\ \omega(\psi) & \psi < 0 \end{cases} \qquad (2.42b)$$

is formally (1.26), we may use a variational approach and consider varying (2.42a) over a suitable class of functions ψ satisfying any necessary fixed boundary conditions, but taking the integration over the fixed domain including both the rotational ($\psi < 0$) and irrotational ($\psi > 0$) parts of the flow. Equally we could try to find a stationary value of either term in the integral (2.42a) subject to the other being held constant. The technical details become complicated for problems in both vortex flows [112] and plasma confinement [235]. Here we simply note that if we compute the first variation of (2.42a) having first subdivided the flow region by the curve $f(x,y) = 0$ on which $\psi = 0$, the free boundary contributions are

$$\oint\{[\frac{\partial\psi}{\partial n}]\delta\psi - [\tfrac{1}{2}|\nabla\psi|^2 - \Omega(\psi)]\delta f/|\nabla f|\}ds. \qquad (2.43)$$

Here $\delta\psi$ and δf are the variations in ψ, f and, as usual, [] denotes the jump across the free boundary. Hence, if we allow $\delta\psi$ and δf to vary independently and Ω is continuous at $\psi = 0$, so are ψ and $\partial\psi/\partial n$ in accordance with the discussion after (1.26).

(b) Obstacle Problems

Although the obstacle problem (1.9) for a string is very elementary, its formulation as a variational problem leads to one of the most important ideas in this book. Formally, a concise way of writing (1.9) on the fixed domain AB is a "linear complementarity problem".

$$\frac{d^2u}{dx^2} = 0 \quad \text{or} \quad u - \psi = 0 \quad \text{i.e.} \quad \frac{d^2u}{dx^2}(u - \psi) = 0 \qquad (2.44a)$$

with

$$u - \psi \geq 0, \quad \frac{d^2u}{dx^2} \leq 0. \tag{2.44b}$$

(2.44) is true everywhere along the string but in addition the physical model requires that u and du/dx are continuous at A', B' i.e.

$$[u]_{A'} = [\frac{du}{dx}]_{A'} = [u]_{B'} = [\frac{du}{dx}]_{B'} = 0 . \tag{2.45}$$

A second fixed domain approach is to regard (1.9) as a variational problem. If A', B' were known a priori, then (1.9) would be equivalent to minimising the energy functional $\int_A^{A'} + \int_{B'}^{B} (du/dx)^2 \, dx$ over a class of suitably smooth functions prescribed at A, A', B, B'. Now we restrict $u \geq \psi$ in AB, i.e. we consider

$$\min_{v \geq \psi} \int_A^B (\frac{dv}{dx})^2 \, dx, \tag{2.46}$$

where v is still suitably smooth but is only prescribed at A and B, and enquire whether (2.46) is equivalent to our original problem (1.9). Minimisations such as (2.46) have been called "unilateral" variational problems [107].

Whereas (1.9) is clearly equivalent to (2.44, 2.45), it is the link between (2.46) and (2.44, 2.45) which we wish to establish and which gives rise to the idea of a "variational inequality". We first note that if u and v are candidates for (2.46) then so is $\lambda v + (1-\lambda)u$ for any λ such that $0 \leq \lambda \leq 1$. Thus if, u minimises (2.46),

$$\int_A^B (\frac{du}{dx})^2 \, dx \leq \int_A^B \{(1-\lambda)\frac{du}{dx} + \lambda\frac{dv}{dx}\}^2 \, dx$$

$$= \int_A^B (\frac{du}{dx})^2 \, dx + 2\lambda \int_A^B \frac{du}{dx} (\frac{dv}{dx} - \frac{du}{dx}) \, dx + 0(\lambda^2) \tag{2.47}$$

for small positive λ. Hence, if we define

$$a(u,v-u) = \int_A^B \frac{du}{dx}\left(\frac{dv}{dx} - \frac{du}{dx}\right) dx, \tag{2.48}$$

then

$$a(u,v-u) \geq 0 \tag{2.49}$$

for all suitable v. The converse argument is trivial since $a(u-v,u-v) \geq 0$ so that (2.49) is equivalent to (2.46) and it is called a variational inequality.

Before relating (2.49) to (2.44, 2.45), we note that in the absence of the obstacle (2.49) holds for $v < \psi$ as well as $v \geq \psi$. Thus, writing $v-u = \eta$,

$$a(u,\eta) \geq 0$$

for all η which vanish at A and B. Hence $a(u,\eta) = 0$ for all such η and, by the usual arguments of the calculus of variations, $d^2u/dx^2 = 0$. This Euler equation holds everywhere for the problem where there is no obstacle but it only holds away from the obstacle in the unilateral problem.

The inequality (2.49) is so important because of the simplicity of demonstrating its equivalence with (2.44, 2.45). This equivalence follows because

$$a(u,v-u) = -\left\{\int_A^{A'} + \int_{B'}^{B} \frac{d^2u}{dx^2}(v-u)\,dx\right\} - \int_{A'}^{B'} \frac{d^2\psi}{dx^2}(v-u)\,dx$$

$$+ \left[\frac{du}{dx}(v-u)\right]_A^{A'} + \left[\frac{du}{dx}(v-u)\right]_{B'}^{B} + \left[\frac{d\psi}{dx}(v-u)\right]_{A'}^{B'} .$$

Now from (2.45), $du/dx = d\psi/dx$ at A' and B' and so, if u satisfies (2.44) as well,

$$a(u,v-u) = -\int_{A'}^{B'} \frac{d^2\psi}{dx^2}(v-u)\,dx \geq 0$$

since $u = \psi$, $v \geq \psi$ and $d^2\psi/dx^2 \leq 0$ at the obstacle A'B'. Thus, by using the variational inequality (2.49), we have a plausible argument for the equivalence of (2.44, 2.45) and (2.46), and the latter provides a fixed domain formulation of the problem in a form which is convenient for computation.

These arguments will be made rigorous in Chapter IV, where the numerical procedure will also be explained but let us now consider a similarly intuitive approach to some of the other problems of Chapter I. The membrane obstacle problem (1.10) can be reformulated using a simple generalisation of the above ideas and defining

$$a(u,v) = \iint \nabla u.\nabla v \, dxdy \tag{2.50a}$$

$$(u,v) = \iint uv \, dxdy \tag{2.50b}$$

where the region of integration is the whole membrane; (2.49) then becomes

$$a(u,v-u) \geq (f,v-u). \tag{2.51}$$

for all suitable $v \geq \psi$. Equally double obstacle problems, where obstacles lie on both sides of strings or membranes lead to (2.49) or (2.51) where now $\psi_2 \geq v \geq \psi_1$.

(c) Elastic-plastic Torsion Problem

A very similar argument applies to the torsion problem (1.12, 1.13), which has the complementarity formulation

$$(\nabla^2\phi + 2G\theta)(|\nabla\phi|^2 - k^2) = 0 \tag{2.52a}$$

with

$$\nabla^2\phi + 2G\theta \geq 0, \quad |\nabla\phi|^2 \leq k^2. \tag{2.52b}$$

In the absence of a plastic region, the problem is equivalent to minimising $a(\phi,\phi) - (4G\theta,\phi)$ in the notation of (2.50a,2.50b), the region of integration being the cross section of the cylinder and ϕ being sufficiently smooth and zero on the curved boundary of the cylinder. This suggests the following unilateral variational problem analogous to (2.46); minimise

$$a(v,v) - (4G\theta,v) \tag{2.53}$$

over those v which are suitably smooth, vanish on the surface of the cylinder and in addition are such that $|\nabla v|^2 \leq k^2$ everywhere. In fact there is another unilateral minimisation involving ϕ instead of $\nabla\phi$ [37], but the elliptic variational inequality which establishes the link between (2.52) and (2.53) is for ϕ which is such that

$$a(\phi, v-\phi) \geq (2G\theta, v-\phi) \tag{2.54}$$

for all suitably smooth v vanishing on the cylinder boundary and such that $|\nabla v|^2 \leq k^2$. A numerical implementation of (2.53) is given in [122] and the free boundary between the elastic and plastic regions can be quite complicated geometrically, as we shall see later.

(d) Parabolic Problems

The application of variational ideas to parabolic field equations is less straightforward and less physically motivated than it is to the elliptic equations we have considered so far. Nonetheless there have been at least two distinct variational approaches to parabolic moving boundary problems. The more physically based method is reviewed briefly in [1] but because of its reliance on thermodynamic ideas and the difficulty of systematising the suggested numerical algorithms, we will not describe it further here.

In some cases a more strictly mathematical generalisation of (2.44)-(2.46) to parabolic problems is possible. The oxygen-consumption problem (1.20)-(1.22), with $\gamma = 1$ for simplicity, may be formulated as the following linear complementarity problem:

$$\left(\frac{\partial c}{\partial t} - \frac{\partial^2 c}{\partial x^2} + R\right) c = 0 \quad \text{in} \quad 0 < x < d, \quad d > s(t) \tag{2.55a}$$

with

$$c \geq 0, \ \frac{\partial c}{\partial t} - \frac{\partial^2 c}{\partial x^2} + R \geq 0 \tag{2.55b}$$

together with the continuity conditions that c and $\partial c/\partial x$ are continuous. Although there is no obvious physical analogue of (2.46) or (2.53), there is a generalisation of (2.49) and (2.54), namely

$$\left(\frac{\partial c}{\partial t}, v-c\right) + a(c, v-c) \geq (-R, v-c) \tag{2.56}$$

for all suitably smooth v satisfying appropriate conditions at $x = 0, d$. The notation is as in (2.50), the integrals being over $0 \leq x \leq d$. (2.56) is called a parabolic variational inequality and it may be solved numerically by a time-stepping procedure in which a certain quadratic functional is minimised at each time step. The details will be explained in Chapter IV, where a comparison will also be made with the truncation method.

(e) Problems with discontinuous derivatives at the unknown boundary

All the free and moving boundary problems considered above have involved second order differential field equations and in the formulation of these problems as "complementarity problems", continuity of the first derivatives of the dependent variable has been required. This continuity is crucial in the steps relating (2.44, 2.45) to (2.49) in the obstacle problem and it would be lacking if, for example, we were to attempt a fixed domain solution of the Hele-Shaw problem (1.28)-(1.31) by defining $p \equiv 0$ in the region unoccupied by liquid. The development of devices which convert such problems into ones with continuous first derivatives have created much interest in variational inequalities recently. Such devices are usually called "Baiocchi transformations", [19] , although the transformations themselves have been used before in other contexts, [216]. They are typified by the following analysis of the Hele-Shaw problem for an expanding liquid region; the "smoothing" transformation we use seems to have no obvious physical interpretation, but it will be put into a more general mathematical setting in Chapter VI.

Let us define the moving boundary $f(x,y,t) = 0$ enclosing the liquid region Ω for $t \geq 0$ in the form $t = \ell(x,y)$ so that (1.30) becomes

$$\nabla p. \nabla \ell = -1. \tag{2.57}$$

For points (x,y) which are traversed by the moving boundary, define

$$U(x,y,t) = \begin{cases} \int_{\ell}^{t} p(x,y,\tau)\,d\tau, & t > \ell \\ 0 & , t \leq \ell \end{cases} \tag{2.58a}$$

and let

$$U(x,y,t) = \int_{0}^{t} p(x,y,\tau)\,d\tau \tag{2.58b}$$

for points within the region Ω_0 occupied by liquid at $t = 0$.

The function U is useful because not only does it satisfy the simple field equations

$$\nabla^2 U = 0 \quad \text{in} \quad \Omega_0 \tag{2.59a}$$

$$\nabla^2 U = 1 \quad \text{in} \quad \Omega - \Omega_0, \tag{2.59b}$$

as can be verified by differentiation of (2.58) and use of (2.57), but also

$$U = \frac{\partial U}{\partial n} = 0 \tag{2.60}$$

at the free boundary of Ω at any time t. Hence, when we extend U to be zero outside Ω, we obtain the following linear complementarity problem on a fixed domain sufficiently large to always contain the liquid region:

$$(\nabla^2 U + \chi)\, U = 0 \tag{2.61a}$$

with

$$\nabla^2 U + \chi \leq 0, \quad U \geq 0, \tag{2.61b}$$

where

$$\chi = \begin{cases} 0 & \text{inside } \Omega_0 \\ -1 & \text{outside } \Omega_0. \end{cases} \tag{2.61c}$$

We may now repeat the earlier manipulations for the obstacle problem to show for example that when liquid is being injected at a constant flow rate $\partial p/\partial n = Q$ across a prescribed boundary Γ_0, U satisfies the variational inequality [97]

$$a(U,\ v-U) \geq (\chi,\ v-U) + \langle Qt,\ v-U\rangle\ . \tag{2.62a}$$

Here the notation is as in (2.50), the region of integration being between Γ_0 and the outer boundary of the fixed domain and

$$\langle Qt,\ v-U\rangle = t \oint_{\Gamma_0} Q(v-U)\,ds \tag{2.62b}$$

and v is any sufficiently smooth non-negative function. Thus we can finally show that U solves the unilateral variational problem

$$\min_{v\geq 0}\{\tfrac{1}{2}a(v,v) - (\chi,v) - \langle Qt,\ v\rangle\}. \tag{2.63}$$

We have already noticed that there are differences between Hele-Shaw flows with expanding and contracting liquid regions and another difference is reflected in (2.59, 2.60). If, for a contracting liquid, we defined $p \equiv 0$ for $t > \ell$ and tried to compute

$$U(x,y,t) = \int_{\ell}^{t} p(x,y,\tau)\,d\tau \tag{2.58c}$$

for points for which $\ell(x,y)$ is defined, we would find that $U(x,y,0)$ would satisfy the Poisson equation (2.59b) locally near the free boundary $\partial\Omega_0$ together with the homogeneous Cauchy data (2.60) on $\partial\Omega_0$. Thus $U(x,y,0)$ would satisfy an ill-posed problem and even if $\partial\Omega_0$ was an analytic curve, no global solution of this problem would exist in general. We will see other aspects of the differences between the problems of expanding and contracting liquid regions in Chapter V.

We have remarked in Chapter I that there is an analogy between steady flow through a porous medium under gravity and "travelling-wave" solutions of the Hele-Shaw problem. The famous problem of a rectangular dam between two reservoirs (Fig.9) may be modelled by

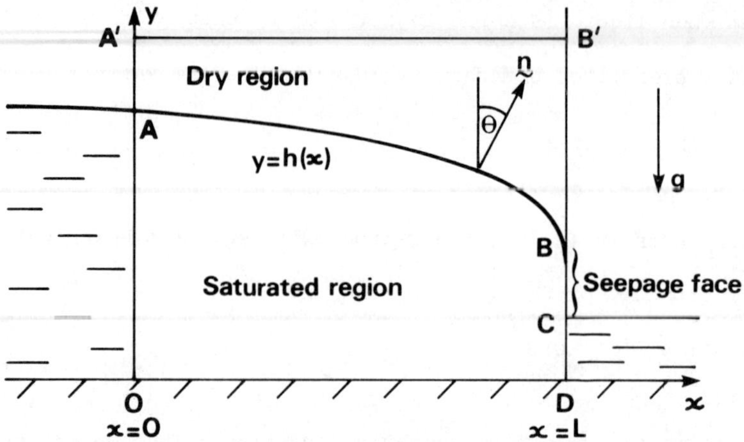

Fig. 9. Rectangular Dam.

$$\nabla^2 p = 0, \quad 0 < x < L, \quad 0 < y < h(x) \tag{2.64}$$

where the pressure p is a prescribed hydrostatic pressure at the reservoirs OA, CD, the flow $\partial p/\partial y + \rho g = 0$ across the impermeable base OD, $p = 0$ at the seepage face BC and the free boundary conditions (1.36, 1.37) are

$$p = 0 = \frac{\partial p}{\partial n} + \rho g \cos \theta \quad \text{at } y = h(x) \tag{2.65a}$$

so that

$$\frac{\partial p}{\partial y} - \frac{dh}{dx}\frac{\partial p}{\partial x} + \rho g = 0 \text{ there.} \tag{2.65b}$$

On physical grounds we expect that this model will predict outflow, $\partial p/\partial x \leq 0$, on BC and indeed this inequality is a consequence of the maximum principle. The situation is more complicated for seepage faces on non-rectangular dams as we shall see in Chapter VI.

Motivated by (2.58a) for a travelling-wave solution moving with constant speed in the y-direction, we define

$$U(x,y) = \begin{cases} \int_y^{h(x)} p(x,y')\,dy', & y < h(x) \\ 0 \quad , & y > h(x) \end{cases} \tag{2.66}$$

so that by differentiation and using (2.65b)

$$\nabla^2 U = \begin{cases} \rho g , & y < h(x) \\ 0 , & y > h(x) \end{cases} \qquad (2.67)$$

with

$$U = \frac{\partial U}{\partial n} = 0 \quad \text{at} \quad y = h(x). \qquad (2.68)$$

By defining U to be zero in the dry part of the dam, the way is now open for a variational inequality formulation, and this will be described in detail in Chapter IV. Although, as mentioned earlier, this particular dam problem can be reduced to a transcendental equation using conformal mapping, the mathematical theory of much more general problems can be treated by transformations related to (2.66)-(2.68), and this will also be described later.

4. SUMMARY

We have indicated several different methods for reformulating and analysing free and moving boundary problems of the type described in Chapter I. Each has its own advantages and disadvantages, and we have not made any real attempt to assess these here. For example we have made no mention of the difficulties associated with implementing minimisations such as (2.46). We have already remarked that a sensible discretisation of the primitive model usually leads to results which are acceptable physically, but it is often difficult to justify such discretisations as far as convergence and stability are concerned. It is often equally difficult to apply a method for which theoretical justification is available to seemingly unsuitable primitive models.

In this chapter we have only succeeded in finding weak or variational formulations of problems which are in some sense conservation laws or minimum energy statements. Nonetheless, we believe that these are the two methods which at present give the best compromise between practical applicability and theoretical justification. Our aim in the next three chapters is to justify this belief with at least an outline of the mathematical theory behind the methods, together with some practical examples.

III The Enthalpy Method for the Stefan Problem

In this chapter we will consider the precise mathematical formulation and numerical solution of the two-phase Stefan problem. To do this, we will need several definitions and ideas from functional analysis and for reasons of clarity we will list and explain these in the appendix. References to the appendix will be denoted by (A).

1. WEAK FORMULATION OF THE STEFAN PROBLEM

Let Ω be an open bounded region in Euclidean space $\mathbb{R}^N$ lying between two fixed closed smooth surfaces $\partial\Omega^-$ and $\partial\Omega^+$ (Fig. 10). For any T > 0, let $\Omega_T = \Omega \times (0,T)$. We suppose that the phase boundary Γ may be described in

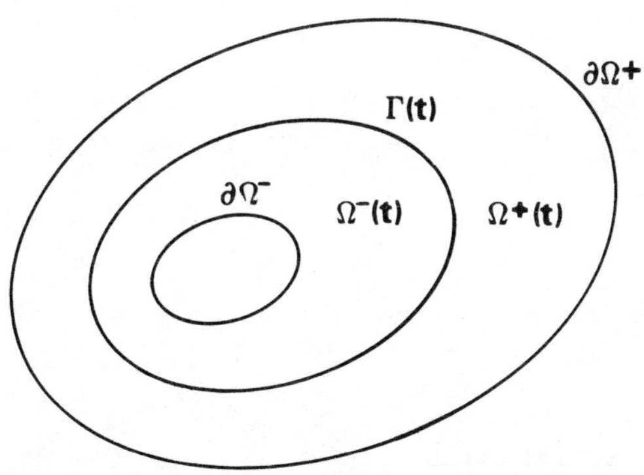

Fig. 10. Stefan Problem.

terms of a smooth function $S(\underset{\sim}{x},t)$ so that we may define

$$\Gamma = \{(\underset{\sim}{x},t) \in \Omega_T : S(\underset{\sim}{x},t) = 0\} \tag{3.1a}$$

$$\Gamma(t) = \{\underset{\sim}{x} \in \Omega : S(\underset{\sim}{x},t) = 0\} \tag{3.1b}$$

$$\Omega^+(t) = \{\underset{\sim}{x} \in \Omega: S(\underset{\sim}{x},t) > 0 \text{ and } u(\underset{\sim}{x},t) > 0\} \text{ (liquid)} \tag{3.1c}$$

$$\Omega^-(t) = \{\underset{\sim}{x} \in \Omega: S(\underset{\sim}{x},t) < 0 \text{ and } u(\underset{\sim}{x},t) < 0\} \text{ (solid)} \tag{3.1d}$$

$$\Omega_T^+ = \bigcup_{t<T} \Omega^+(t), \quad \Omega_T^- = \bigcup_{t<T} \Omega^-(t) \tag{3.1e}$$

$$\Omega = \Omega^+(t) \cup \Omega^-(t) \cup \Gamma(t). \tag{3.1f}$$

With the formulation (2.33) in mind, the generalised enthalpy is defined as a monotone mapping $\beta : \mathbb{R} \to \mathbb{R}$ which is continuously differentiable away from the origin where it has a finite jump discontinuity. Thus

$$\beta'(r) \geq \alpha > 0 \text{ for } r \neq 0; \qquad [\beta]_{0-}^{0+} = L > 0 \tag{3.2a,b}$$

where, as usual, square brackets denote jumps and the density is suppressed by being taken to be unity. Then, for given enthalpy and temperature distributions $H(\underset{\sim}{x},t)$ and $u(\underset{\sim}{x},t)$

$$H(\underset{\sim}{x},t) \in \beta(u(\underset{\sim}{x},t)) \tag{3.3}$$

almost everywhere in Ω_T, where we have adopted the convention that (3.3) means

$$u \neq 0, \qquad H = \beta(u) \tag{3.4a}$$

$$u = 0, \qquad H \in [0,L]. \tag{3.4b}$$

The classical formulation of the two-phase Stefan problem is to find $\{u,S\}$ such that

$$\frac{\partial H}{\partial t} = \nabla^2 u, \quad (\underset{\sim}{x},t) \in \Omega_T \backslash \Gamma \tag{3.5a}$$

$$H = \beta(u), \quad u \neq 0 \tag{3.5b}$$

with, for simplicity, Dirichlet data

$$u = g = \begin{cases} g_1 < 0, & \underset{\sim}{x} \in \partial\Omega^- \\ & \\ g_2 > 0, & \underset{\sim}{x} \in \partial\Omega^+ \end{cases}, \quad t \in (0,T) \tag{3.6}$$

and initial conditions

$$u(\underset{\sim}{x},0) = u_0(\underset{\sim}{x}) \begin{cases} > 0, & \underset{\sim}{x} \in \Omega^+(0) \\ = 0, & \underset{\sim}{x} \in \Gamma(0) \\ < 0, & \underset{\sim}{x} \in \Omega^-(0). \end{cases} \tag{3.7}$$

The energy balance, or "Stefan" condition at Γ is

$$[\nabla u]_{\text{solid}}^{\text{liquid}} . \nabla S = L \frac{\partial S}{\partial t}, \qquad (\underset{\sim}{x},t) \in \Gamma \tag{3.8a}$$

which we may rewrite as

$$[\nabla u]_{\text{solid}}^{\text{liquid}} . \underset{\sim}{\nu}_{\underset{\sim}{x}} = L\nu_t, \ (\underset{\sim}{x},t) \in \Gamma \tag{3.8b}$$

where $\underset{\sim}{\nu} = (\underset{\sim}{\nu}_{\underset{\sim}{x}}, \nu_t)$ is normal to Γ. For simplicity we assume that the initial and boundary data are continuous and compatible. Then the functions $\{u,S\}$ are said to be a classical solution if

(i) u, $\partial S/\partial t$, ∇S are continuous in Ω_T

(ii) $\partial u/\partial t$, ∇u, $\nabla^2 u$ are continuous in $\Omega_T^+ \cup \Omega_T^-$

(iii) (3.5)-(3.8) are satisfied.

The existence of a classical solution to the multi-dimensional two-phase Stefan problem is an open question [210], so, following (2.30), we now proceed to define a weak solution.

Let $H_0(\underset{\sim}{x})$ be the initial enthalpy distribution defined on $\Omega^+(0) \cup \Omega^-(0)$ by (3.7). A weak solution of the Stefan problem is defined to be a pair of bounded integrable functions $\{u,H\}$ defined on Ω_T such that (3.3) is satisfied and the integral identity

$$\int_{\Omega_T} \left(H \frac{\partial \phi}{\partial t} + u\nabla^2\phi\right) d\underset{\sim}{x}dt = - \int_{\Omega} H_0(\underset{\sim}{x})\,\phi(\underset{\sim}{x},0)\,d\underset{\sim}{x} + \int_0^T \int_{\partial\Omega} g \frac{\partial \phi}{\partial n}\, dS_{\underset{\sim}{x}}\, dt \tag{3.9}$$

holds for all test functions $\phi \in F$, where

$$F = \{\phi \in C^1(\bar{\Omega}_T) : \partial^2\phi/\partial x_i \partial x_j \in C(\bar{\Omega}_T); \quad \phi(\underset{\sim}{x},T) = 0, \quad \underset{\sim}{x} \in \Omega;$$
$$\phi(\underset{\sim}{x},t) = 0, \quad \underset{\sim}{x} \in \partial\Omega, \quad t \in (0,T].\} \tag{3.10}$$

Here and subsequently we use the notation $\underset{\sim}{n}$ for the outward pointing unit normal to a surface and $d\underset{\sim}{S}$ for an element of surface area.

Theorem III.1

(i) A classical solution is also a weak solution.

(ii) If $\{u,H\}$ is a weak solution and there is a function S such that (3.1), (3.8) and (i), (ii) on p. 63 hold, then $\{u,S\}$ is a classical solution.

Proof

(i) Let $\phi \in F$. Then, if $\{u,S\}$ is a classical solution

$$0 = \int_{\Omega_T^+} \left(\frac{\partial H}{\partial t} - \nabla^2 u\right)\phi \, d\underset{\sim}{x}dt$$

$$= -\int_{\Omega_T^+} \left(H \frac{\partial \phi}{\partial t} + u\nabla^2\phi\right) d\underset{\sim}{x}dt + \int_{\Omega_T^+} \left(\frac{\partial}{\partial t}(H\phi) + \nabla(u\nabla\phi - \phi\nabla u)\right) d\underset{\sim}{x}dt .$$

Now, by the divergence theorem,

$$\int_{\Omega_T^+} \left(H \frac{\partial \phi}{\partial t} + u\nabla^2\phi\right) d\underset{\sim}{x}dt = \int_{\partial\Omega_T^+} (H\phi,\ u\nabla\phi - \phi\nabla u).(n_t,\underset{\sim}{n}_{\underset{\sim}{x}})\, d\underset{\sim}{S}_{\underset{\sim}{x},t}$$

$$= -\int_{\Omega^+(0)} H_0(\underset{\sim}{x})\phi(\underset{\sim}{x},0)\,d\underset{\sim}{x} + \int_0^T \int_{\partial\Omega^+} g_2 \frac{\partial\phi}{\partial n}\, d\underset{\sim}{S}_{\underset{\sim}{x}}\, dt + \int_\Gamma \phi(-H,\nabla u).(n_t,\underset{\sim}{n}_{\underset{\sim}{x}})\, d\underset{\sim}{S}_{\underset{\sim}{x},t},$$

using the initial condition on H, the final condition on ϕ and the fixed boundary conditions on $\partial\Omega^+$ for ϕ and u. A similar expression holds in Ω_T^- and, since from (3.8b)

$$\int_\Gamma \phi\left(-[H]_{\text{solid}}^{\text{liquid}}\, n_t + [\nabla u]_{\text{solid}}^{\text{liquid}}.\underset{\sim}{n}_{\underset{\sim}{x}}\right) d\underset{\sim}{S}_{\underset{\sim}{x},t} = 0,$$

the classical solution satisfies (3.9).

(ii) Let $\phi \in C_0^\infty(\Omega_T^+)$. Then, substituting into (3.9),

$$0 = \int_{\Omega_T^+} \left(H \frac{\partial\phi}{\partial t} + u\nabla^2\phi\right) d\underset{\sim}{x}dt = - \int_{\Omega_T^+} \left(\frac{\partial H}{\partial t} - \nabla^2 u\right)\phi \; d\underset{\sim}{x}dt.$$

Since u is smooth, this implies that $\partial H/\partial t = \nabla^2 u$ in Ω_T^+, and the same argument holds for Ω_T^-. The boundary conditions satisfied by u are obtained similarly by taking test functions whose supports are neighbourhoods of the surfaces $\partial\Omega^+ \times (0,T)$, $\partial\Omega^- \times (0,T)$ and Γ.

Theorem III.2

When the data is sufficiently smooth, a weak solution exists.

Proof

In [114] it is shown that the data is sufficiently smooth for this theorem to hold if there exists a function G such that $\partial G/\partial t$, ∇G and $\partial^2 G/\partial x_i \partial x_j$ are Hölder continuous on $\overline{\Omega}_T$ with $G = g$ on $\partial\Omega \times [0,T]$ and $G(\underset{\sim}{x},0) = u_0(\underset{\sim}{x})$ on Ω. We will return to [114] shortly, but we first describe the uniqueness proof given in [136].

Uniqueness

Let $\{u_i,H_i\}$, $i = 1,2$, be two weak solutions. Then

$$\int_{\Omega_T} (H_1-H_2)\left(\frac{\partial\phi}{\partial t} + e\,\nabla^2\phi\right) d\underset{\sim}{x}dt = 0, \quad \phi \in F \tag{3.11}$$

where

$$e(\underset{\sim}{x},t) = \begin{cases} \dfrac{u_1(\underset{\sim}{x},t)-u_2(\underset{\sim}{x},t)}{H_1(\underset{\sim}{x},t)-H_2(\underset{\sim}{x},t)} & u_1 \neq u_2 \\ 0 & u_1 = u_2 \end{cases}$$

and

$$H_i(\underset{\sim}{x},t) \in \beta(u_i(\underset{\sim}{x},t)), \quad i = 1,2.$$

Approximating e by a sequence of smooth functions e_m, we define ϕ_m to be the solution of

$$\frac{\partial \phi_m}{\partial t} + c_m \nabla^2 \phi_m = f \quad \text{in } \Omega_T \tag{3.12}$$

with

$$\phi_m = 0 \quad \text{on} \quad \partial\Omega \times (0,T), \quad \phi_m(\underset{\sim}{x},T) = 0$$

and where $f \in C_0^\infty(\Omega_T)$. Since $\phi_m \in F$, (3.11) implies that if we write $I = \int_{\Omega_T} (H_1-H_2) f \, d\underset{\sim}{x}dt$

then

$$I = \int_{\Omega_T} (H_1-H_2)(e_m-e)\nabla^2\phi_m \, d\underset{\sim}{x}dt.$$

Estimates on the solution of (3.12) then imply that $|I| < \varepsilon_m$ where $\varepsilon_m \to 0$ as $n \to \infty$. Hence $I = 0$ and so $H_1(\underset{\sim}{x},t) = H_2(\underset{\sim}{x},t)$ almost everywhere in Ω_T. Since H_i uniquely determines u_i, we also have that $u_1 = u_2$ almost everywhere in Ω_T.

Existence

The existence of a weak solution can be proved by considering a sequence of finite difference approximations $\{u^h, H^h\}$, where h denotes a mesh parameter, [136]. The stability of this sequence, as defined in the sense of certain norms being uniformly bounded independently of h (A; see also p. 81), yields the existence of weakly convergent subsequences whose limits are the weak solution.

The alternative approach [114] mentioned above approximates (3.5) by a sequence of nonlinear parabolic equations which are obtained by smoothing the enthalpy. In this regularisation (p. 30), we let $\beta_\varepsilon : \mathbb{R} \to \mathbb{R}$ be smooth functions which converge uniformly to β away from the origin as $\varepsilon \to 0$. Then, by obtaining estimates for the solution u_ε of

$$\beta'_\varepsilon(u_\varepsilon)\frac{\partial u_\varepsilon}{\partial t} = \nabla^2 u_\varepsilon \tag{3.13}$$

with

$$u_\varepsilon = g \quad \text{on} \quad \partial\Omega$$

$$u_\varepsilon(\underset{\sim}{x},0) = u_o \quad \text{in} \quad \Omega,$$

it can be shown that u_ε converges weakly to the weak solution u in $H^1(\Omega_T)$ and converges strongly to u in $L^2(\Omega_T)$ as $\varepsilon \to 0$, (A).

We note that for any $\phi \in C_0^\infty(\Omega_T)$, $\phi \in F$ and so $\int_{\Omega_T} (H\ \partial\phi/\partial t + u\nabla^2\phi)\, d\underset{\sim}{x}dt = 0$. Hence the distributional derivatives (A) of H and u satisfy

$$\langle\frac{\partial H}{\partial t} - \nabla^2 u,\phi\rangle = 0, \qquad \phi \in C_0^\infty(\Omega_T) \tag{3.14}$$

where the notation $\langle .,. \rangle$ is defined in the appendix, and so the equation

$$\frac{\partial H}{\partial t} = \nabla^2 u$$

holds in the sense of distributions.

We also note that we may define a weak solution of (3.5)-(3.8) as a pair of bounded integrable functions {u,H} such that ∇u is integrable, u = g on $\partial\Omega$ and

$$\int_0^T \{(H, \frac{\partial\phi}{\partial t}) - (\nabla u,\nabla\phi)\}\, dt + (H_0,\phi(\underset{\sim}{x},0)) = 0 \tag{3.15}$$

for all $\phi \in C^\infty(\overline{\Omega}_T) \cap F$, where (.,.) is the L^2 inner product on Ω [84, 133]. Under the assumption that ∇u is integrable, (3.15) is equivalent to (3.9, 3.10) and the fact that ∇u is integrable follows from the remark made at the top of this page. Now we have already mentioned a finite difference discretisation which converges to the solution of (3.9); the finite element discretisation which we will propose in part II of this chapter will converge to the solution of (3.15). The demonstration of this will thus furnish another existence proof of the weak solution of the Stefan problem.

We also mention that the weak solutions (3.9, 3.15) have been studied in a slightly different way using monotonicity methods for evolution equations in Hilbert spaces [33, 84, 149].

The above ideas have been applied to more general Stefan problems than (3.5)-(3.8). In particular, (3.6) can be replaced by a nonlinear condition, [173, 188], which is important because the Kirchoff transformation (2.32) can change linear boundary conditions into nonlinear ones. Also the thermal parameters may be prescribed functions of $(\underset{\sim}{x},t)$ as long as the jump (3.2b) in β at the phase boundary is positive and β is strictly monotonic increasing, [173], and there may be volumetric or body heating, when (3.5a) is replaced by

$$\frac{\partial H}{\partial t} = \nabla^2 u + f(\underset{\sim}{x},t,u), \tag{3.16}$$

[9, 84, 133, 173, 229]. A special case of (3.16) will be discussed in Chapter V. Finally we note that the Hele-Shaw problem (1.28)-(1.31) for an expanding liquid region can be written in the form (3.5) where now

$$H \in \beta(p) = \begin{cases} 1 \;, & p > 0 \\ [0,1], & p = 0 \\ 0 \;, & p < 0 \end{cases} \tag{3.17}$$

so that β is not strictly monotone. Some analytical and numerical implications are discussed in [74, 188].

2. NUMERICAL ANALYSIS OF THE ENTHALPY METHOD

(i) Discretisation of the Weak Formulation

For convenience with the numerical analysis we consider a problem with constant thermal parameters and homogeneous Dirichlet boundary conditions. Thus, in contrast to our usual convention, we take the phase change temperature to be $u = u_F \neq 0$. Because some of the interesting examples involve body heating, we will include it from the beginning, but for ease of exposition, we will take it to be independent of u. Thus we consider

$$\frac{\partial H}{\partial t} = \nabla^2 u + f(\underset{\sim}{x},t) \quad \text{in} \quad \Omega_T \tag{3.18a}$$

where

$$H(\underset{\sim}{x},t) \in \beta(u(\underset{\sim}{x},t)), \qquad (\underset{\sim}{x},t) \in \Omega_T \tag{3.18b}$$

with

$$\beta(r) = \begin{cases} \alpha_2(r-u_F) + L, & r > u_F \\ [0,L] \quad , & r = u_F \\ \alpha_1(r-u_F) \quad , & r < u_F \end{cases} \tag{3.18c}$$

and

$$u(\underset{\sim}{x},t) = 0, \qquad \underset{\sim}{x} \in \partial\Omega \tag{3.19}$$

$$H(\underset{\sim}{x},0) = H_0(\underset{\sim}{x}), \qquad \underset{\sim}{x} \in \Omega. \tag{3.20}$$

We note that if $\alpha = \min(\alpha_1,\alpha_2) > 0$, then

$$(\beta(r)-\beta(q))(r-q) \geq \alpha(r-q)^2 \tag{3.21}$$

for all $r, q \in \mathbb{R}$, which reflects the monotonicity of β.

Finite Difference Approximations

For simplicity we assume that Ω_T may be covered by a uniform rectangular net. Then the finite difference approximation to (3.18a) is obtained by replacing spatial derivatives with central differences and the time derivative with a one-sided difference. The discrete problem is to find vectors $\underset{\sim}{H}^n$ and $\underset{\sim}{u}^n$, the values of enthalpy and temperature at the mesh points at the nth time step, such that

$$(\underset{\sim}{H}^{n+1}-\underset{\sim}{H}^n)/k + A\,\underset{\sim}{u}^{n+\theta} = \underset{\sim}{f}^{n+\theta} \tag{3.22a}$$

and such that the components $(\underset{\sim}{H}^{n+1})_i$, $(\underset{\sim}{u}^{n+1})_i$ satisfy

$$(\underset{\sim}{H}^{n+1})_i \in \beta((\underset{\sim}{u}^{n+1})_i). \tag{3.22b}$$

Here $\theta \in [0,1]$, $\underset{\sim}{u}^{n+\theta} = \theta\,\underset{\sim}{u}^{n+1} + (1-\theta)\underset{\sim}{u}^n$ and similarly for $\underset{\sim}{f}$, k is the time step, - A is the matrix representing the finite difference approximation to the Laplacian on the net, and $\underset{\sim}{H}^0$ is the vector of the given initial enthalpy values. If a mesh point lies on the phase boundary, then the initial enthalpy value at that point may be obtained by a local smoothing procedure as $(\underset{\sim}{H}^0)_i = \int_{\delta v} H_0(\underset{\sim}{x})\,d\underset{\sim}{x}/m(\delta v)$, where $m(\delta v)$ is the measure of a small region δv surrounding the mesh point. A similar procedure may be necessary when f is not smooth.

The values $\theta = 0, \frac{1}{2}$ and 1 correspond to the explicit, Crank-Nicolson and fully implicit time discretisations respectively. Simple examples are, in one dimension

$$\Omega \equiv (0,1), \quad h = 1/m, \; i \in [0,m] \tag{3.23a}$$

$$(H_i^{n+1}-H_i^n)/k + (2u_i-u_{i-1}-u_{i+1})^{n+\theta}/h^2 = f_i^{n+\theta}, \quad i \in (0,m) \tag{3.23b}$$

$$u_0 = u_m = 0 \tag{3.23c}$$

and, in two dimensions,

$$\Omega \equiv (0,1) \times (0,1), \quad h = 1/m, \quad i,\, j \in [0,m] \tag{3.24a}$$

$$(H_{i,j}^{n+1}-H_{i,j}^n)/k + (4u_{i,j}-u_{i-1,j}-u_{i+1,j}-u_{i,j+1}-u_{i,j-1})^{n+\theta}/h^2 = f_{i,j}^{n+\theta} \tag{3.24b}$$

$$u_{i,0} = u_{i,m} = u_{0,j} = u_{m,j} = 0, \quad i,j \in (0,m). \tag{3.24c}$$

<u>A Finite Element Approximation</u>

Let Ω be a polygonal region in $\mathbb{R}^2$. Then Ω may be covered by a union of triangles with disjoint interiors T^h, where h denotes the length of the longest triangle edge, and each triangle edge is either the edge of another triangle or has end points on $\partial\Omega$. We shall assume that the smallest angle in the triangulation is bounded below, independently of h, by θ_0 and that the ratio of the largest element diameter to the smallest element diameter is bounded above, also independently of h, by γ. Let V^h denote the space of continuous functions which are linear on each triangle and which vanish on $\partial\Omega$. For any continuous function v on Ω, we define the interpolate v_I by $v_I \in V^h$ and

$v_I = v$ at each vertex of T^h. Then, for each $v \in H_0^1(\Omega) \cap H^2(\Omega)$, (A), the interpolation estimate

$$|v-v_I|_0 + h|v-v_I|_1 \leq c\, h^2 \tag{3.25}$$

holds, [57], where c is a positive constant and

$$|v|_0 = \{\int v^2 d\underset{\sim}{x}\}^{\frac{1}{2}}, \qquad |v|_1 = \{\int |\nabla v|^2 d\underset{\sim}{x}\}^{\frac{1}{2}} .$$

Any $v \in V^h$ may be written as $v = \sum_{j\in J} v_j b_j(\underset{\sim}{x})$ where J is the set of interior nodes, v_j are the nodal values and $b_j(\underset{\sim}{x})$ is a piecewise linear basis function satisfying $b_j(\underset{\sim}{x}_i) = \delta_{ij}$, where $\underset{\sim}{x}_i$ is the position of node i. The L^2 inner product (.,.) is replaced on V^h by the discrete L^2 inner product $(.,.)_h$ defined by the "lumped" integration rule

$$(u,v)_h = \frac{1}{3} \sum_{j\in J} W_j\, u_j v_j = \underset{\sim}{u}^T M\underset{\sim}{v} \tag{3.26}$$

where $\underset{\sim}{u} = (u_i)$, W_j is the sum of the areas of elements attached to node j and the positive diagonal matrix M is called the lumped mass matrix. For this integration rule, it can be shown that there exist positive constants c_i, independent of h, such that for all u, $v \in V^h$

$$|v|_0 \leq |v|_h \leq c_0|v|_0 \tag{3.27a}$$

where

$$|v|_h = \{(v,v)_h\}^{\frac{1}{2}},$$

and

$$|(u,v)-(u,v)_h| \leq c_1 h^2 |u|_1\ |v|_1 \leq c_2 h|u|_0\ |v|_1, \tag{3.27b}$$

[58]. The discrete inverse norm inequality

$$|v|_1 \leq S(h)\,|v|_h \tag{3.28}$$

holds for all $v \in V^h$, where

$$S(h) = \sqrt{\frac{3}{2}}\,\frac{h}{\min A(T)} \tag{3.29a}$$

when the triangulation has all acute-angled triangles, and

$$S(h) = \frac{3}{2}\frac{h}{\min A(T)} \tag{3.29b}$$

otherwise, and A(T) denotes the area of a triangle T. An upper bound for S(h) is given by

$$\sin\theta_0 . hS(h)/\gamma^2 \le \begin{cases} \sqrt{6} & (3.30a) \\ 3 & (3.30b) \end{cases}$$

for each of the triangulations in (3.29a, b) respectively.

If we again use the notation $a(u,v) = \int_\Omega \nabla u . \nabla v d\underset{\sim}{x}$, then a weak formulation of (3.18a) when β is *smooth* is

$$(\frac{\partial H}{\partial t}, v) + a(u,v) = (f,v) \tag{3.31}$$

for all $v \in H_0^1(\Omega)$ and this motivates the following finite element discretisation. Using the same notation as in (3.22), find H^n and u^n such that $u^n \in V^h$ and

$$((H^{n+1}-H^n)/k,v)_h + a(u^{n+\theta},v) = (f^{n+\theta},v)_h \tag{3.32}$$

for all $v \in V^h$, where $H^n = \sum_i H_i^n b_i(\underset{\sim}{x})$, $u^n = \sum_i u_i^n b_i(\underset{\sim}{x})$, $f^n = \sum_i f(\underset{\sim}{x}_i, t^n) b_i(\underset{\sim}{x})$ and

$$H_i^n \in \beta(u_i^n). \tag{3.33}$$

Taking $v = b_i(\underset{\sim}{x})$ for each interior node i, we obtain

$$M_{ii}(H_i^{n+1}-H_i^n)/k + \sum_j a(b_i,b_j)u_j^{n+\theta} = M_{ii} f_i^{n+\theta}.$$

Hence, denoting the interior nodal values by $\underset{\sim}{H}$, $\underset{\sim}{u}$ and $\underset{\sim}{f}$,

$$(\underset{\sim}{H}^{n+1}-\underset{\sim}{H}^n)/k + A\,\underset{\sim}{u}^{n+\theta} = \underset{\sim}{f}^{n+\theta} \tag{3.34}$$

where -A, the finite element discretisation of the Laplacian on Ω with Dirichlet boundary conditions for the given triangulation, is defined by

$$A = M^{-1}\hat{A}, \quad \hat{A}_{ij} = a(b_i,b_j), \quad i,j \in J. \tag{3.35}$$

A is a sparse positive definite symmetric matrix.

Construction of the Discrete Data for the Finite Element Approximation

We shall see later (section (v)) that, for convergence, the approximations to the initial data must satisfy, (A),

$$H_h^0 \rightharpoonup H_0 \quad \text{in } L^2(\Omega) \tag{3.36a}$$

$$u_h^0 \rightharpoonup u_0 \quad \text{in } H_0^1(\Omega) \tag{3.36b}$$

as $h \to 0$, where $(H_h^0)_i \in \beta((u_h^0)_i)$

and the notation is as in (3.33). Also a mesh function $f_{h,k}(t)$ must be constructed from $f^{n+\theta}$ and satisfy

$$f_{h,k}(t) \rightharpoonup f \quad \text{in } L^2(\Omega_T) \tag{3.36c}$$

and

$$\|f_{h,k}\|_{L^2(\Omega_T)} \leq c$$

where the constant c is independent of h,k. However, there is no explicit computational need for u_h^0 to be defined when $\theta = 1$.

Difficulties arise when the data H_0, u_0 and f are not smooth and in particular are not defined pointwise. For example, there might initially be a sharp interface across which the enthalpy jumps, or the body heating may have a finite jump discontinuity across a surface in Ω_T. Two possible constructions for the initial data are:-

a) Let the initial enthalpy distribution $H_0 \in L^\infty(\Omega)$ and be smooth in a finite number of subdomains Ω_j such that $\bar{\Omega} = \bigcup_j \bar{\Omega}_j$, and only have jump discontinuities across the boundaries $\partial\Omega_j$. Further, suppose $u_0 \in W^{1,\infty}(\Omega)$, (A). Then it is sufficient to define

$$u_h^0 = \sum_i u_0(\underset{\sim}{x}_i)\, \underset{\sim}{b}_i(\underset{\sim}{x}) \tag{3.37a}$$

$$H_h^0 = \sum_i H_i^0 b_i(\underset{\sim}{x}) \quad \text{where } H_i^0 \begin{cases} = \beta(u_0(\underset{\sim}{x}_i)),\ u_0(\underset{\sim}{x}_i) \neq u_F \\ \in \beta(u_F),\ u_0(\underset{\sim}{x}_i) = u_F . \end{cases} \tag{3.37b}$$

(3.37a) is possible because u_0 is continuous.

b) Suppose $u_0 \in H_0^1(\Omega)$ and $u_0 \neq u_F$ almost everywhere in Ω. Then it is sufficient to define, [58],

$$a(u_h^0, v) = a(u_0, v) \tag{3.38}$$

for all $v \in V^h$, and H_h^0 as in (3.37b).

If $f(\underset{\sim}{x},t)$ is not defined pointwise, then it is sufficient to take

$$f^{n+\theta} \in V^h, \qquad (f^{n+\theta}, v)_h = \frac{1}{k} \int_{nk}^{(n+1)k} (f(\underset{\sim}{x},t), v)\,dt$$

for all $v \in V^h$ for (3.36c) to hold.

We have described some possible discretisations of the weak solution of the Stefan problem, and it remains to consider their existence, uniqueness, stability and convergence properties. Implicit time stepping schemes with a discontinuous enthalpy were first analysed in [58] and our development of the finite element approximation to the weak solution of the Stefan problem follows that of [58, 93]. An implicit scheme for regularised problems with a continuous enthalpy, where the phase change takes place over a small temperature range, is considered in [162].

(ii) Existence and Uniqueness of the Solution of the Discrete Problem

The explicit schemes for (3.22a) or (3.34) are

$$\underset{\sim}{H}^{n+1} = \underset{\sim}{H}^n + k\underset{\sim}{f}^n - kA\underset{\sim}{u}^n \tag{3.39}$$

and define $\underset{\sim}{H}^{n+1}$ by a straightforward calculation of a vector. In particular, the one-dimensional finite difference scheme is

$$H_i^{n+1} = H_i^n + kf_i^n + \frac{k}{h^2}(u_{i-1} - 2u_i + u_{i+1})^n \tag{3.40}$$

and, since β is invertible,

$$u_i^{n+1} = \begin{cases} u_F + (H_i^{n+1}-L)/\alpha_2, & H_i^{n+1} > L \\ u_F & , \quad H_i^{n+1} \in [0,L] \\ u_F + H_i^{n+1}/\alpha_1 & , \quad H_i^{n+1} < 0 . \end{cases} \tag{3.41}$$

If, as in the Hele-Shaw problem, $\alpha_1 = \alpha_2 = 0$, then H does not define u uniquely and an implicit scheme must be used, [74].

Implicit schemes for either finite difference or finite element discretisations lead to the system of nonlinear algebraic equations

$$\underset{\sim}{H} + C\,\underset{\sim}{u} = \underset{\sim}{b}, \quad H_i \in \beta(u_i), \tag{3.42}$$

where C is a symmetric positive definite matrix. If β was a continuous monotone function, then $\underset{\sim}{H} + C\,\underset{\sim}{u} - \underset{\sim}{b}$ would be the gradient of a functional and the existence of a solution of (3.42) would follow from optimisation theory. However, by generalising the concept of a derivative we will show that solving (3.42) is equivalent to finding the minimum value of a non-differentiable convex functional.

Let $\mathcal{H}$ be a Hilbert space with inner product <.,.>. Then a functional $\Phi: \mathcal{H} \to \mathbb{R}$ is said to be convex if

$$\Phi(\lambda x_1 + (1-\lambda)x_2) \le \lambda\Phi(x_1) + (1-\lambda)\Phi(x_2) \tag{3.43}$$

for all $\lambda \in [0,1]$ and all $x_1, x_2 \in \mathcal{H}$. Φ is strictly convex if strict inequality holds in (3.43) for $x_1 \neq x_2$. The concept of derivative is generalised to non-differentiable convex functionals by introducing the subdifferential $\partial\Phi(.)$ defined by

$$\partial\Phi(x) = \{z \in \mathcal{H}: \Phi(w) - \Phi(x) \ge \langle z, w-x\rangle \text{ for all } w \in \mathcal{H}\} \tag{3.44}$$

and any element of $\partial\Phi(x)$ is called a subgradient of Φ at x.

For example, let $\phi : \mathbb{R} \to \mathbb{R}$ be the continuous convex piecewise quadratic function

$$\phi(r) = \tfrac{1}{2}\,\alpha_2(r-u_F)_+^2 + \tfrac{1}{2}\,\alpha_1(r-u_F)_-^2 + L(r-u_F)_+ \tag{3.45}$$

where $r_+ = \begin{cases} r, & r > 0 \\ 0, & r < 0 \end{cases}$, $r_- = \begin{cases} 0, & r > 0 \\ r, & r < 0 \end{cases}$.

Then $\partial\phi(r) = \beta(r)$ as defined by (3.18c) and this is depicted in Fig. 11.

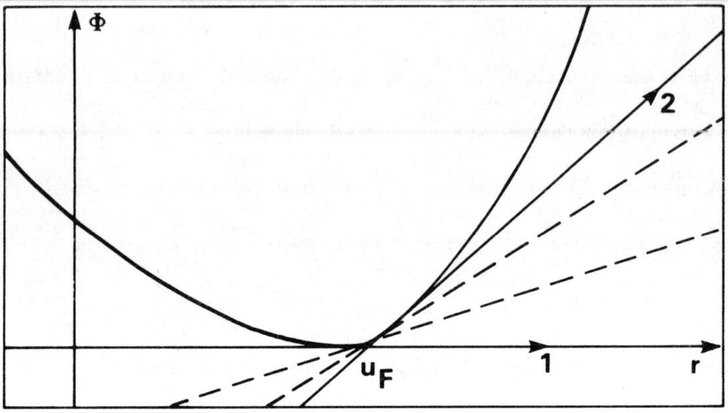

Fig. 11. Subgradient of (3.45).

The gradients of the dashed lines lying between the lines 2 and 1, which are the tangents at $r = u_F$ to the functions $\frac{1}{2}\alpha_2(r-u_F)^2 + L(r-u_F)$ and $\frac{1}{2}\alpha_1(r-u_F)^2$ respectively, are elements of $\partial\phi(u_F)$. When $\Phi(.)$ is differentiable at x, then $\partial\Phi(x)$ consists of a single element, namely the derivative evaluated at x. Geometrically a differentiable convex functional lies above its tangent at a point, whereas a nondifferentiable convex functional lies above the set of hyperplanes defined by the subdifferential. This suggests that $0 \in \partial\Phi(x)$ is a sufficient condition for Φ to have a minimum at x, and indeed this follows immediately from the definition (3.44). The following theorem is a result of optimisation theory, [220].

Theorem III.3

Let Φ be a strictly convex continuous functional on a Hilbert space H such that $\Phi(v) \to \infty$ as $\|v\| \to \infty$. Then there exists a unique minimum of Φ which is characterised by $0 \in \partial\Phi(u)$.

We now consider (3.42) and let H be the Euclidean space $\mathbb{R}^J$ and define $\Phi(\underset{\sim}{v})$ by

$$\Phi(\underset{\sim}{v}) = \sum_{j=1}^{J} \phi(v_j) + \tfrac{1}{2}\langle C\underset{\sim}{v},\underset{\sim}{v}\rangle - \langle \underset{\sim}{b},\underset{\sim}{v}\rangle \tag{3.46}$$

where ϕ is given by (3.45) and $\langle \underset{\sim}{x},\underset{\sim}{y}\rangle = \underset{\sim}{x}^T\underset{\sim}{y}$. Now it is easy to check that a finite sum of convex functionals is convex. Moreover, the quadratic form $\tfrac{1}{2}\langle C\underset{\sim}{v},\underset{\sim}{v}\rangle - \langle \underset{\sim}{b},\underset{\sim}{v}\rangle$ is strictly convex since C is positive definite and, since $\langle C\underset{\sim}{v},\underset{\sim}{v}\rangle \geq \lambda_m\langle \underset{\sim}{v},\underset{\sim}{v}\rangle$, where λ_m is the smallest eigenvalue of C, and $\phi(v_j) \geq 0$, Φ satisfies the hypotheses of theorem III.3. Moreover, a straightforward calculation shows that the subdifferential of (3.46) is such that

$$\underset{\sim}{v} \in \partial\Phi(\underset{\sim}{u}) \Leftrightarrow v_i \in \partial\phi(u_i) + (C\underset{\sim}{u}-\underset{\sim}{b})_i = \beta(u_i) + (C\underset{\sim}{u}-\underset{\sim}{b})_i . \tag{3.47}$$

Hence, by theorem III.3, (3.42) has a unique solution which also solves

$$\Phi(u) = \inf_{v\in H} \Phi(v) .$$

(iii) <u>A Successive Overrelaxation Algorithm</u>

(3.42) may be solved by an iterative method. The classical result for successive point relaxation when β is a continuous monotone function is that convergence is assured when the relaxation parameter ≤ 1, [185]. It has been observed, [161],, that if overrelaxation is used at points changing phase in an iteration, then oscillation about the phase change temperature may occur. Hence, it has been suggested, [162], that when β is single-valued, the Gauss-Seidel scheme should be used at such points. The following algorithm for the multivalued β (3.18c) was suggested and analysed in [93]. In describing the algorithm, just in this section we take $u_F = 0$.

Given an initial guess $\underset{\sim}{u}_0$ for the temperature and a relaxation parameter $\omega \in (0,2)$, $\underset{\sim}{u}^{k+1}$ is obtained from $\underset{\sim}{u}^k$ by successively calculating, component by component,

a) the solution of the inclusion

$$z_i^k \equiv b_i - \sum_{j<i} C_{ij} u_j^{k+1} - \sum_{j>i} C_{ij} u_j^k \in C_{ii} v_i^{k+1} + \beta(v_i^{k+1}) \tag{3.48}$$

which is

$$v_i^{k+1} = \begin{cases} (z_i^k - L)/(C_{ii} + \alpha_2), & z_i^k > L \\ 0, & z_i^k \in [0,L] \\ z_i^k/(C_{ii} + \alpha_1), & z_i^k < 0, \end{cases} \tag{3.49}$$

and then testing

b) if $u_i^k\{u_i^k + \omega(v_i^{k+1} - u_i^k)\} > 0$ (3.50)

then

$$u_i^{k+1} = u_i^k + \omega(v_i^{k+1} - u_i^k),$$

otherwise

$$u_i^{k+1} = v_i^{k+1}.$$

Theorem III.4, [93]

(i) For any initial guess $\underset{\sim}{u}_0$, the iteration (3.48)-(3.50) is convergent.

(ii) If the solution to (3.42) satisfies

$$u_i = 0 \Leftrightarrow (\underset{\sim}{b} - C\underset{\sim}{u})_i \in (0,L)$$

then there exists a k_0, depending on the data, such that $\{u_i^k > 0 \text{ if } u_i > 0\}$, $\{u_i^k < 0 \text{ if } u_i < 0\}$ and $\{u_i^k = 0 \text{ if } u_i = 0\}$ for all $k > k_0$.

The theorem is proved by showing that each iteration decreases Φ and then convergence follows from compactness arguments (A). If the condition in (ii) holds, then after a finite number of iterations, the set of zero mesh points has been determined and so the method reduces to the usual SOR method for solving linear equations. This suggests that the optimum value for ω is that which minimises the spectral radius of the iteration matrix for the mesh points at which u is non-zero. This is discussed further in Chapter IV.

(iv) Stability

We say that a discretisation method for a nonlinear parabolic equation is stable if certain norms of the discrete solution are uniformly bounded independently of the mesh parameters. In this section, it is convenient to use the notation $v = \underset{\sim}{v}$, $\langle u,v\rangle = \underset{\sim}{u}^T M\underset{\sim}{v}$, $|v|^2 = \langle v,v\rangle$, $\|v\|^2 = \underset{\sim}{v}^T MA\underset{\sim}{v}$ and $\delta_k v^n = (v^{n+1}-v^n)/k$, with the convention that $M = I$ for the finite difference approximation.

The norm inequality

$$\|v\| \le S(h)\,|v| \tag{3.51}$$

holds, where $S(h) = 2\sqrt{N}/h$ for finite differences on a uniform mesh in N dimensions, $N = 1,2$, and $S(h)$ is given by (3.29) or (3.30) for the finite element approximation. In the latter case, we have the identities $\|v\| \equiv |v|_1$, $|v| \equiv |v|_h$.

Taking the inner product of

$$\delta_k H^n + Au^{n+\theta} = f^{n+\theta} \tag{3.52}$$

with $k\delta_k u^n$ gives

$$k\langle\delta_k H^n, \delta_k u^n\rangle + \langle Au^{n+\theta}, u^{n+1}-u^n\rangle = k\langle f^{n+\theta}, \delta_k u^n\rangle. \tag{3.53}$$

Now, from the monotonicity of β, (3.21),

$$\begin{aligned} k^2 \langle \delta_k H^n, \delta_k u^n\rangle &= \sum_i (\beta(u_i^{n+1}) - \beta(u_i^n))(u_i^{n+1}-u_i^n)M_{ii} \\ &\ge \alpha \sum_i (u_i^{n+1}-u_i^n)^2 M_{ii} \\ &= \alpha k^2 |\delta_k u^n|^2 \end{aligned} \tag{3.54}$$

and, rewriting $\langle Au^{n+\theta}, u^{n+1}-u^n\rangle$, we obtain

$$\alpha k|\delta_k u^n|^2 + \tfrac{1}{2}\{(2\theta-1)\|u^{n+1}-u^n\|^2 + \|u^{n+1}\|^2 - \|u^n\|^2\} \le k\langle f^{n+\theta}, \delta_k u^n\rangle. \tag{3.55}$$

Then, using the Cauchy-Schwartz inequality $\langle u,v\rangle \le |u|\,|v|$, and also the fact that $2ab \le \varepsilon a^2 + b^2/\varepsilon$ for any $\varepsilon > 0$, we may sum to obtain

$$2\alpha \sum_{j=0}^{n} k|\delta_k u^j|^2 + \|u^{n+1}\|^2 \le \|u^0\|^2 + \frac{k}{\varepsilon}\sum_{j=0}^{n} |f^{j+\theta}|^2 + \varepsilon k \sum_{j=0}^{n} |\delta_k u^j|^2$$

$$+ (1-2\theta)\sum_{j=0}^{n} \|u^{j+1}-u^j\|^2 . \qquad (3.56)$$

We can now prove

Theorem III.5

Providing the stability condition

$$(1-2\theta)k\{S(h)\}^2 < 2\alpha \qquad (3.57)$$

holds when $\theta < \frac{1}{2}$, there is a constant C independent of h,k, such that

$$\alpha \sum_{j=0}^{n} k|\delta_k u^j|^2 + \|u^{n+1}\|^2 \le C\{\|u^0\|^2 + \sum_{j=0}^{n} k|f^{j+\theta}|^2\} . \qquad (3.58)$$

Proof

(i) <u>$\theta \ge \frac{1}{2}$</u>.

Since $1-2\theta \le 0$, we may take $\varepsilon = \alpha$ in (3.56) to obtain

$$\alpha \sum_{j=0}^{n} k|\delta_k u^j|^2 + \|u^{n+1}\|^2 \le \|u^0\|^2 + \frac{1}{\alpha}\sum_{j=0}^{n} k|f^{j+\theta}|^2.$$

(ii) <u>$\theta < \frac{1}{2}$</u>.

Applying (3.51) to the last term of (3.56), we obtain

$$(2\alpha-\varepsilon-(1-2\theta)k\{S(h)\}^2)\sum_{j=0}^{n} k|\delta_k u^j|^2 + \|u^{n+1}\|^2 \le \|u^0\|^2 + \frac{1}{\varepsilon}\sum_{j=0}^{n} k|f^{j+\theta}|^2.$$

We note that the stability restriction (3.57) is that which holds for the finite difference or finite element approximation of the linear parabolic equation $\alpha\ \partial u/\partial t = \nabla^2 u + f$, [198].

(v) <u>Convergence of the Finite Element Discretisation</u>

The convergence of explicit finite difference schemes has been discussed in

[9, 136]. To consider the finite element discretisation, we define the following mesh functions on [0,T):

$$u_{h,k}(t) = u^n, \quad u^\theta_{h,k}(t) = u^{n+\theta}, \quad H_{h,k}(t) = H^n \tag{3.59a}$$

$$f_{h,k}(t) = f^{n+\theta}, \quad \hat{u}_{h,k}(t) = \frac{(n+1)k-t}{k}u^n + \frac{t-nk}{k}u^{n+1} \tag{3.59b}$$

for $t \in [nk,(n+1)k)$. The stability theorem III.5 and (3.36a, b, c) imply that the sequences (A)

a) $\{u_{h,k}\}, \{u^\theta_{h,k}\}$ and $\{\hat{u}_{h,k}\}$ are uniformly bounded in $L^\infty(0,T;H^1_0(\Omega))$.

b) $\{\hat{u}_{h,k}\}$ is uniformly bounded in $H^1(\Omega_T)$.

c) $\{H_{h,k}\}$ is uniformly bounded in $L^\infty(0,T;L^2(\Omega))$.

The uniform boundedness of $H_{h,k}(t)$ is a consequence of the uniform boundedness of $\|u^n\|$ since $|\beta(r)| \le \max(\alpha_1,\alpha_2)(r-u_F)+L$. This implies $|H^n|_h \le C_1|u^n|_h + C_2$, where C_i are independent of h, and by Poincarés inequality (A), $|u^n|_0 \le C|u^n|_1$, where C is a positive constant depending on Ω.

Hence, by weak compactness (A), there exist subsequences, which we do not distinguish notationally from the original sequences, such that

$$\hat{u}_{h,k} \rightharpoonup u \text{ in } L^2(0,T;H^1_0(\Omega)) \tag{3.60a}$$

$$u_{h,k} \rightharpoonup u \text{ in } L^2(0,T;H^1_0(\Omega)) \tag{3.60b}$$

$$u^\theta_{h,k} \rightharpoonup u \text{ in } L^2(0,T;H^1_0(\Omega)) \tag{3.60c}$$

$$\hat{u}_{h,k} \to u \text{ in } L^2(\Omega_T) \tag{3.60d}$$

$$H_{h,k} \rightharpoonup H \text{ in } L^2(\Omega_T). \tag{3.60e}$$

The strong convergence (3.60d) follows from the compactness result that bounded sets in $H^1(\Omega_T)$ are contained in compact subsets of $L^2(\Omega_T)$, [222],(A). .

Now for any $v \in C^\infty(\Omega_T) \cap F$, the mesh functions $v_{h,k}(t)$ and $v'_{h,k}(t)$ defined on [0,T) by (p. 70)

$$v_{h,k}(t) = (v(nk))_I \equiv v^n, \quad t \in [nk,(n+1)k)$$

$$v'_{h,k}(t) = (v^{n+1}-v^n)/k, \quad t \in [nk,(n+1)k)$$

have the approximation properties

$$t = nk, \qquad v_I(t) \to v(t) \text{ in } H^1_0(\Omega)$$

$$v_{h,k} \to v \text{ in } L^2(0,T;H^1_0(\Omega))$$

$$v'_{h,k} \to \frac{\partial v}{\partial t} \text{ in } L^2(0,T;H^1_0(\Omega)).$$

Substituting $v = v_I(nk)$ into (3.32) and summing yields

$$0 = k \sum_{n=0}^{N-1} ((H^{n+1}-H^n)/k,v^n)_h + a(u^{n+\theta},v^n) - (f^{n+\theta},v)_h$$

$$= (H^N,v^{N-1})_h - (H^0,v^0)_h + k \sum_{n=1}^{N-1} (-H^n,(v^n-v^{n-1})/k)_h$$

$$+ k \sum_{n=0}^{N-1} a(u^{n+\theta},v^n) - (f^{n+\theta},v^n)_h$$

$$= (H_{h,k}(T),\ v_{h,k}(T-k))_h - (H_h^0,\ v_{h,k}(0))_h$$

$$+ \int_k^T - (H_{h,k}(t),\ v'_{h,k}(t-k))_h dt + \int_0^T \{a(u^\theta_{h,k}(t),\ v_{h,k}(t))$$

$$- (f_{h,k}(t),\ v_{h,k}(t))_h\}dt.$$

We now let $h,k \to 0$ and use the strong convergence properties of $v_{h,k}$ and $v'_{h,k}$, the weak convergence of the mesh functions (3.60), the approximation of the data (3.36) and the numerical integration estimate (3.27a, b) to obtain

$$0 = -(H_0,v(0)) + \int_0^T \{(-H, \frac{\partial v}{\partial t}) + a(u,v) - (f,v)\}dt$$

which is (3.15) when body heating is included. Thus $\{u,H\}$ is a weak solution provided we can show that $H(x,t) \in \beta(u(x,t))$ almost everywhere in Ω_T.

In order to demonstrate this, we let $\zeta(.)$ be the Lipschitz continuous inverse of $\beta(.)$ so that $u = \zeta(H) \Leftrightarrow H \in \beta(u)$. Then, for $v \in C^\infty(\Omega_T) \cap F$,

$$\sum_{n=0}^{N-1} k(u^n-\zeta(v^n),\ H^n-v^n)_h \geq 0$$

where we recall $v^n = (v(nk))_I$ and, as $h,k \to 0$, by (3.60d, e),

$$\int_0^T (u-\zeta(v),\ H-v)\ dt \geq 0\ . \tag{3.61}$$

In (3.61) we may replace v by a sequence v^i converging in $L^2(\Omega_T)$ to $\lambda w + (1-\lambda)H$, $\lambda \in [0,1]$, to show that

$$\int_0^T (u-\zeta(\lambda w+(1-\lambda)H),\ H-w)\ dt \geq 0.$$

Hence, writing $w = H \pm z$, where $z \in L^2(\Omega_T)$ and letting $\lambda \to 0$, we find

$$\int_0^T (u-\zeta(H),z)\ dt = 0 \quad \text{for } z \in L^2(\Omega_T),$$

and thus $u = \zeta(H)$ almost everywhere in Ω_T.

Since the weak solution is unique, the limits are independent of the subsequences and so the sequences themselves converge.

The implication (3.61) $\Rightarrow$ $H \in \beta(u)$ is a consequence of $\beta(.)$ being a maximal monotone mapping, [35].

The following theorem has now been proved.

Theorem III.6

Let $f \in L^2(\Omega_T)$, $H_0 \in L^2(\Omega)$ and $\zeta(H_0) \in H_0^1(\Omega)$, with $\zeta(H_0) \neq u_F$ almost everywhere in Ω. Then if (3.57) holds, there exists a weak solution to the Stefan problem to which the finite element approximation converges in the sense that

$$u_{h,k} \to u \quad \text{in } L^2(\Omega_T)$$

$$H_{h,k} \rightharpoonup H \quad \text{in } L^2(\Omega_T)$$

$$u_{h,k} \rightharpoonup u \text{ in } L^2(0,T;\ H_0^1(\Omega)).$$

(vi) Numerical Examples

(a) We consider the freezing of a liquid with zero freezing temperature which is initially in the region $x > 0$ at a positive temperature. A constant negative temperature is applied at $x = 0$ for $t > 0$. Allowing different constant conductivities in each phase and selecting suitable units, the problem may be written classically as

$$\frac{\partial u}{\partial t} = \frac{\partial^2 u}{\partial x^2}, \quad u < 0, \quad x < s(t) \tag{3.62a}$$

$$\frac{\partial u}{\partial t} = \kappa \frac{\partial^2 u}{\partial x^2}, \quad u > 0, \quad x > s(t) \tag{3.62b}$$

$$\kappa \left.\frac{\partial u}{\partial x}\right|_{\text{liquid}} - \left.\frac{\partial u}{\partial x}\right|_{\text{solid}} = -L\frac{ds}{dt}, \quad u = 0, \quad x = s(t) \tag{3.62c}$$

with initial and boundary conditions

$$u(x,0) = u^* > 0, \quad 0 < x < \infty \tag{3.63a}$$

$$u(0,t) = -u_0 < 0, \quad u(\infty,t) = u^*, \quad t > 0. \tag{3.63b}$$

The explicit Neumann solution is, [50],

$$s = 2at^{\frac{1}{2}} \tag{3.64a}$$

where

$$u_0 e^{-a^2}/\mathrm{erf}a = \kappa^{\frac{1}{2}} u^* e^{-a^2/\kappa}/\mathrm{erfc}(a/\kappa^{\frac{1}{2}}) = aL\pi^{\frac{1}{2}} \tag{3.64b}$$

and

$$u_{\text{solid}} = -u_0(1 - \mathrm{erf}(x/2t^{\frac{1}{2}})/\mathrm{erf}a) \tag{3.64c}$$

$$u_{\text{liquid}} = u^*(1 - \mathrm{erfc}(x/2\kappa^{\frac{1}{2}}t^{\frac{1}{2}})/\mathrm{erfc}(a/\kappa^{\frac{1}{2}})). \tag{3.64d}$$

Considering the case of water and ice, we take $\kappa = 1.44/11.5$ and $L = 73.6 \times 11.5/5.3$, and solve on the interval $(0,1)$ from time $t = 0.1$ to $t = 4.0$, taking the initial and boundary conditions from (3.64). The two parameters defining the problem are chosen to be $u_0 = 20$, $a = 0.5/6^{\frac{1}{2}}$, with u^* satisfying (3.64b).

A new temperature $\bar{u}$ is given by the Kirchoff transformation

$$\bar{u} = \begin{cases} u, & u < 0 \\ \kappa u, & u > 0 \end{cases} \tag{3.65}$$

so that

$$\frac{\partial H}{\partial t} = \frac{\partial^2 \bar{u}}{\partial x^2} \tag{3.66}$$

in the sense of distributions (A), where

$$H = \begin{cases} \bar{u} \quad , & \bar{u} < 0 \\ \bar{u}/\kappa + L, & \bar{u} > 0 \quad . \end{cases} \tag{3.67}$$

Discretising in the usual way, (3.23) gives

$$H_i^{n+1} = H_i^n + k(\bar{u}_{i+1} + \bar{u}_{i-1} - 2\bar{u}_i)^{n+\theta}/h^2, \quad i \in (0,M) \tag{3.68}$$

with $\bar{u}_0^n$ and $\bar{u}_M^n$ given, $h = 1/M$, $k = 4/N$. For $\theta = 0$, stability requires the restriction, (3.57), $k/h^2 \le \frac{1}{2}$, i.e. $N \ge 8M^2$ since $\alpha = \min(1,\kappa^{-1})$. Taking $h = 1/20$ implies that the maximum stable explicit time step is $k = 1/800$, and in practice, using these mesh sizes in the explicit scheme gives approximately the same relative error as that of the fully implicit scheme with the same h but with a time step $k = 1/20$. This means that the implicit scheme is at least twice as fast as the explicit. We find that the cases $\theta = 1$ and $\theta = \frac{1}{2}$ are equivalent in speed and accuracy for the mesh sizes used.

When solving the implicit, nonlinear equations, the relaxation factor ω was chosen on the basis of the linear theory of S.O.R. (see [242]) which gives the optimum factor associated with the matrix

$$cI + \frac{\theta k}{h^2} A, \tag{3.69}$$

where A is tridiagonal and given by (3.68), as

$$\omega = 2/(1+(1-p^2)^{\frac{1}{2}}), \quad p = 2\theta k \cos(\pi h)/ch^2. \tag{3.70}$$

Note that (3.69) yields a matrix $c(\bar{u}) + \theta k/h^2 A$, where $c(\bar{u})$ is a diagonal matrix with ith entry 1 or $1/\kappa$ depending on the sign of $\bar{u}_i$. Thus we expect that near the solution $\bar{u}^{n+1}$, we are solving linear equations with matrix $c(\bar{u}^{n+1}) + \theta k/h^2 A$. ω is calculated from (3.70) taking c to be $z + (1-z)/\kappa$ where z is the proportion of mesh points at which $\bar{u}^n$ is negative. This strategy works better in practice than taking a fixed ω in (1.0, 1.1 ..., 1.9).

In Table 1, the relative discrete L^2 errors, $\{\Sigma_i(u_i-u(ih))^2/\Sigma_i u_i^2\}^{\frac{1}{2}}$ are tabulated at certain times for the computations made with h = k and θ = 1. Pointwise convergence is demonstrated in Table 2, where the values of u are tabulated. The analytical solution (3.64) is compared graphically with the computed solution in Figs. 12, 13, the discrete points being plotted from the computed solution. Fig. 12 is a plot of the temperature distribution at different times, while Fig. 13 records the temperature history at x = 0.5.

TABLE 1

% *Discrete L^2 Errors*

t \ h	1/10	1/20	1/40	1/80
0.5	4.48	1.65	1.46	0.27
1.0	1.42	1.01	0.84	0.34
1.5	1.20	0.33	0.37	0.07
2.0	6.26	2.26	0.29	0.21
2.5	9.54	1.23	0.93	0.44
3.0	0.80	0.83	0.90	0.51
3.5	8.44	2.06	1.31	0.12
4.0	2.27	2.38	1.18	0.43

TABLE 2

Computed and Exact Temperature Values

x	t \ h	1/20	1/40	1/80	Exact
0.4	0.5	16.49	16.06	16.35	16.43
	1.0	0.00	0.00	-0.47	-0.39
	2.0	-5.63	-6.04	-6.03	-6.04
	3.0	-8.55	-8.54	-8.55	-8.58
	4.0	-9.99	-10.07	-10.09	-10.10
0.6	0.5	34.00	33.88	34.01	34.13
	1.0	19.54	19.25	19.33	19.35
	2.0	2.05	1.97	2.00	1.90
	3.0	-2.85	-2.84	-2.93	-2.96
	4.0	-4.99	-5.22	-5.18	-5.21
0.8	0.5	40.80	40.88	40.98	41.08
	1.0	32.09	31.96	32.01	32.03
	2.0	16.63	16.38	16.44	16.43
	3.0	6.23	6.25	6.19	6.16
	4.0	0.0	-0.56	-0.30	-0.39

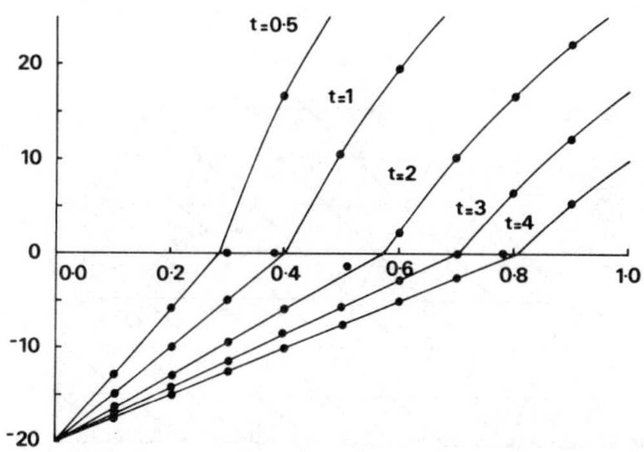

Fig. 12

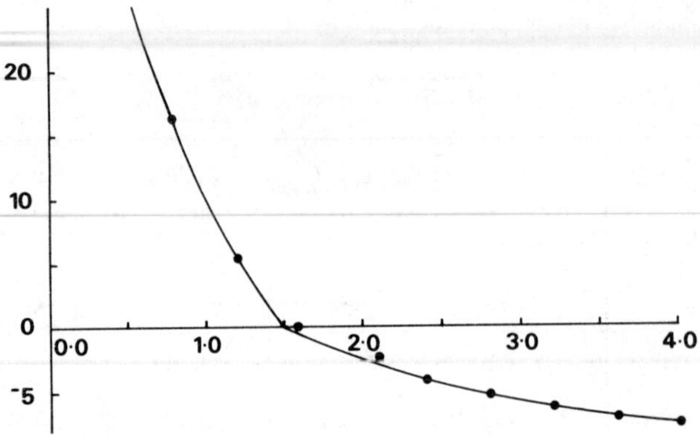

Fig. 13

(b) Let

$$\frac{\partial H}{\partial t} = \frac{\partial^2 u}{\partial x^2} + \frac{\partial^2 u}{\partial y^2} + f, \quad x,y \in \Omega = \{r^2 = x^2 + y^2 < 1\} \tag{3.71}$$

where

$$H = \begin{cases} u, & u < 0 \\ 4u+2, & u > 0 \end{cases} \qquad f = \begin{cases} 8(2e^{-2t}-1) & r > e^{-t} \\ 2(2e^{-2t}-2) & r < e^{-t} \end{cases} \tag{3.72a,b}$$

with $u(x,y,0) = x^2+y^2-1$.

The solution is

$$u_{liquid} = 2(x^2+y^2-e^{-2t}) \tag{3.74}$$

$$u_{solid} = x^2 + y^2 - e^{-2t}. \tag{3.75}$$

The problem is solved over a sector with radial arms $y = 0$, $y = x$, using a triangulation of the type shown in Fig. 14.

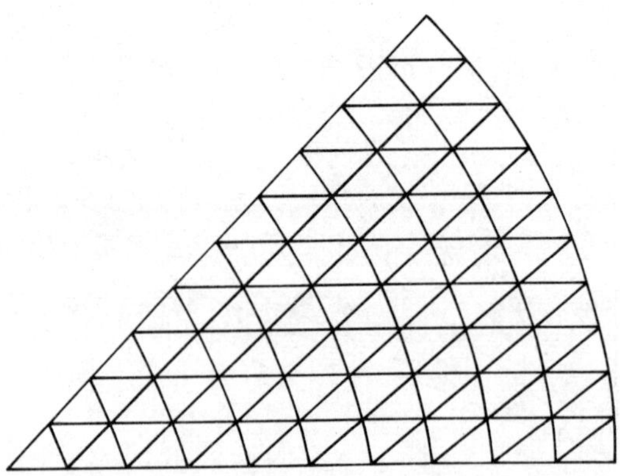

Fig. 14

The time interval is (0,1) and the initial and boundary values are taken from the exact solution together with zero flux conditions on the radial arms. For this triangulation, the explicit scheme is limited in time step by the stability condition (3.57)

$$k \leq h^2/6 \tag{3.76}$$

where h is the mesh spacing along the radius $y = 0$. As in example (a), this restriction on k results in the implicit scheme being more efficient than the explicit one. Taking $h = 1/10$ requires the explicit time step to be no greater than 1/600 and similar accuracy can be attained with $\theta = 1$ and $k = 1/25$ with a gain in speed. The exact solution is compared graphically in Figs. 15, 16 with the solution computed using $h = 1/20$, $k = 1/100$ and $\theta = 1$. Again the discrete points are taken from the numerical solution. Fig. 15 shows the temperature along $y = 0$ at various times and Fig. 16 records the temperature history at two points on $y = 0$. The numerical temperature histories are plotted as continuous curves in Fig. 17. Step-like behaviour near the phase change is apparent, and there is a delay in passing through the phase change value.

For one-dimensional problems a method for obtaining more accuracy for the temperature and position of the moving boundary is given in [245]. It is based on the assumption that the moving boundary crosses a mesh point when the enthalpy at that point is L/2. Interpolation is used to find the time at which this occurs. The corresponding interpolated temperature values are found to be more accurate than at the time levels on the original grid.

Other application of the enthalpy method to Stefan-like problems are given in [20, 32, 66, 75, 217, 247, 253].

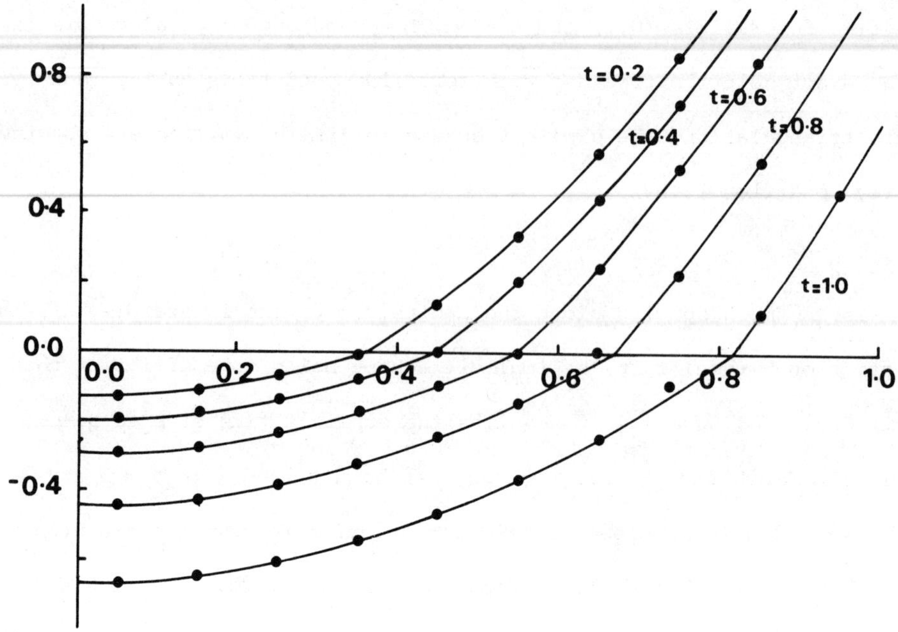
0·8
0·4
0·0
-0·4
0·0
0·2
0·4
0·6
0·8
1·0
t=0·2
t=0·4
t=0·6
t=0·8
t=1·0

Fig. 15.

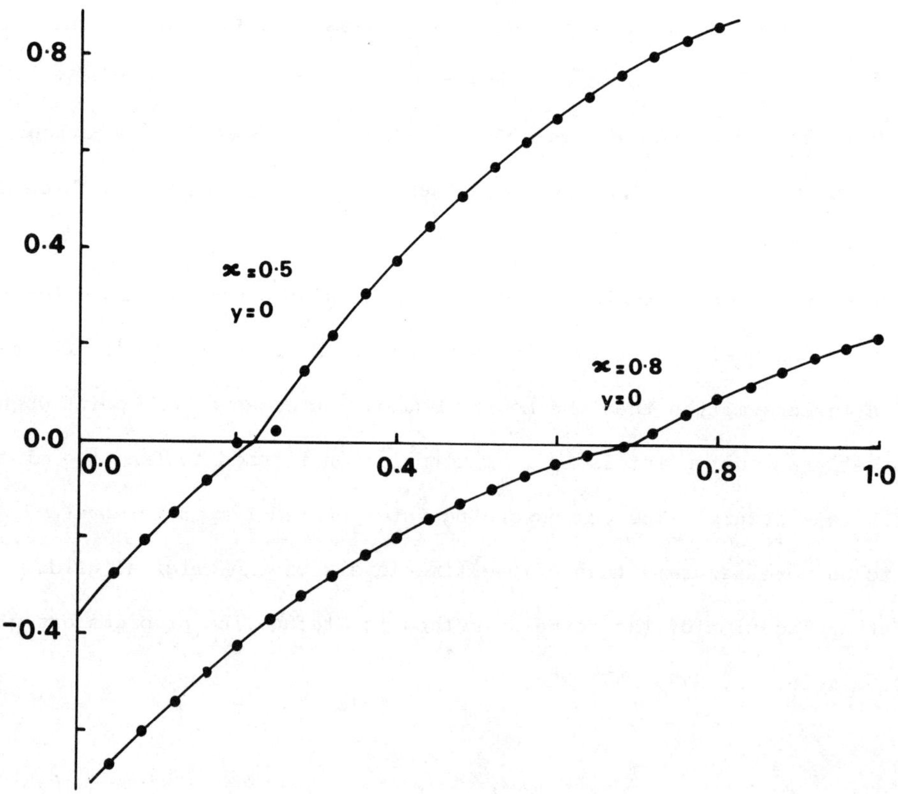
0·8
0·4
0·0
-0·4
0·0
0·4
0·8
1·0
x=0·5
y=0
x=0·8
y=0

Fig. 16.

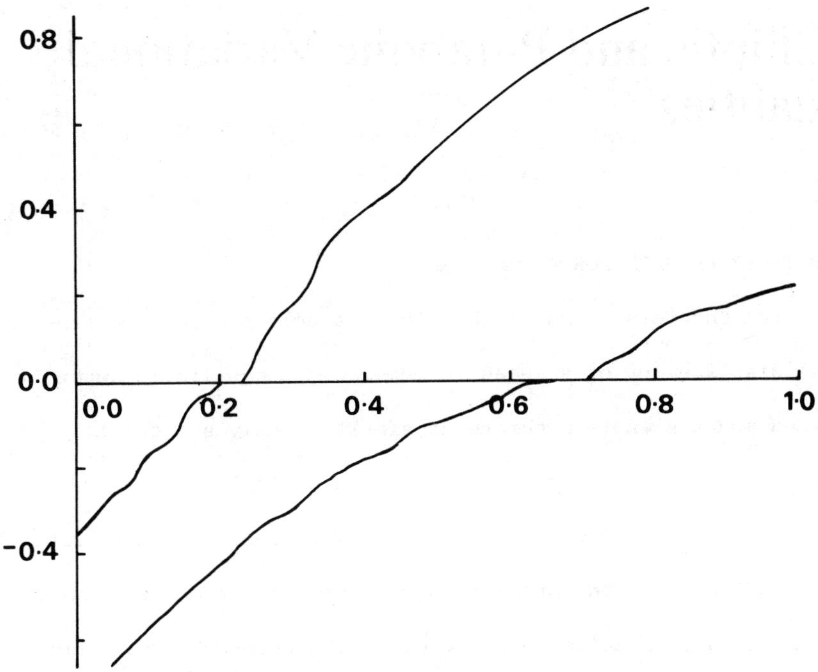

Fig. 17 .

IV Elliptic and Parabolic Variational Inequalities

1. THE OBSTACLE PROBLEM FOR A MEMBRANE

We recall from Chapters I and II that the obstacle problem concerns the small transverse displacement of an elastic membrane in equilibrium when its boundary is fixed and the whole membrane is simultaneously subject to a pressure and constrained to lie on one side of a rigid obstacle. Let Ω denote the open bounded region in $\mathbb{R}^2$ occupied by the undisturbed membrane, whose boundary is $\partial\Omega$, and let f be the pressure and u and ψ the transverse displacements of the membrane and obstacle respectively. The membrane cross section is depicted schematically in Fig. 3 when f = 0. Let the contact region in Ω be denoted by Ω^0 and let Ω^+ be the region in Ω where there is no contact. Then, as in Chapter I,

$$-\nabla^2 u = f, \qquad (x,y) \in \Omega^+, \tag{4.1}$$

$$u(x,y) = \psi(x,y), \qquad (x,y) \in \Omega^0 \tag{4.2}$$

and we assume for simplicity that

$$u = 0, \qquad (x,y) \in \partial\Omega. \tag{4.3}$$

Also, since the membrane slope is continuous,

$$u = \psi, \qquad \nabla u = \nabla\psi, \qquad (x,y) \in \Gamma \tag{4.4}$$

where Γ is the free boundary separating Ω^+ and Ω^0.

In the typical situation when $\psi < 0$ on $\partial\Omega$, $\partial\Omega^0$ does not intersect $\partial\Omega$ and a plan view of the membrane may be as in Fig. 18. .

(4.1)-(4.4) define a free boundary problem for $\{u,\Gamma\}$. However the example below shows that these equations may not have a unique solution unless they are supplemented by the inequalities

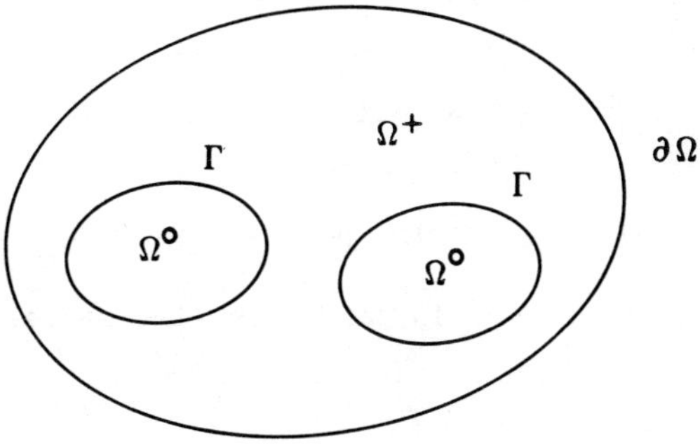

Fig. 18. Plan view of membrane.

$$u \geq \psi, \qquad (x,y) \in \Omega \tag{4.5}$$

and

$$-\nabla^2 u \geq f, \qquad (x,y) \in \Omega, \tag{4.6}$$

where (4.5) is the constraint that the membrane lies on one side of the obstacle and, in Ω^0, (4.6) follows from a force balance when contact is maintained with a rigid obstacle.

Example

Let $\Omega = [-1,1]$, $\psi = -1/18$ and

$$f = \begin{cases} -1 \ , & |x| > \frac{1}{4} \\ 1-32x^2, & |x| \leq \frac{1}{4} \end{cases} .$$

Then two solutions to (4.1)-(4.5) are

(i) $\Gamma = \pm\ 2/3$; $\Omega^+ = \{2/3 < |x| < 1\}$; $\Omega^0 = \{|x| < 2/3\}$

and, in Ω^+, $u = -\ 1/18 + \frac{1}{2}(x \pm 2/3)^2$.

(ii) $\Gamma = \pm\ 2/3,\ \pm\ 1/3$; $\Omega^+ = \{2/3 < |x| < 1\} \cup \{|x| < 1/3\} = \Omega_1^+ \cup \Omega_2^+$,

$$\Omega^0 = \{1/3 < |x| < 2/3\}$$

and

$$u = \begin{cases} -\frac{1}{18} + \frac{1}{2}(x \pm \frac{2}{3})^2, & x \in \Omega_1^+ \\ -\frac{1}{32} + \frac{8x^2}{3}(x^2 - \frac{3}{16}), & |x| < \frac{1}{4} \\ -\frac{1}{18} + \frac{1}{2}(x \pm \frac{1}{3})^2, & \frac{1}{4} < |x| < \frac{1}{3} \end{cases}$$

These solutions are sketched in Fig. 19, and only (ii) satisfies (4.6)

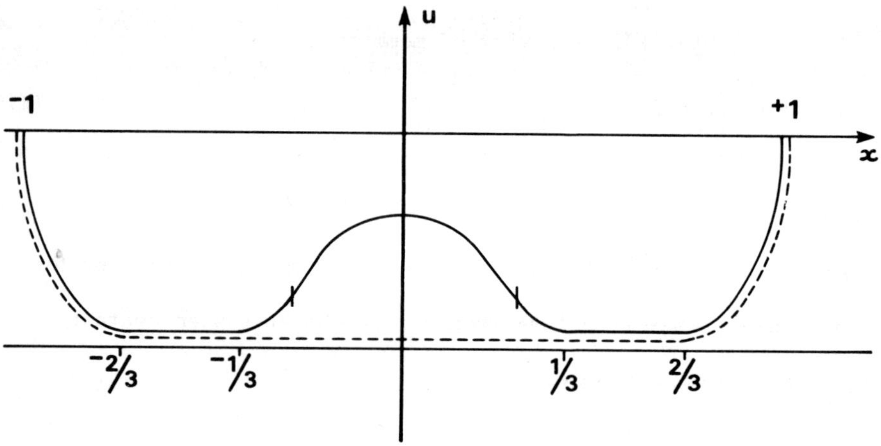

Fig. 19. Non-unique solutions (i)-----, (ii) ———.

The elastic strain energy for any displacement field $v(x,y)$ is

$$E(v) = \int_\Omega (\tfrac{1}{2}\{(\frac{\partial v}{\partial x})^2 + (\frac{\partial v}{\partial y})^2\} - fv)\,dxdy. \tag{4.7}$$

Letting V be the set of displacement fields for which $E(v)$ is finite, the argument outlined in (2.46)-(2.51) suggests that the membrane displacement minimizes E over the set of admissible displacement fields

$$K = \{v \in V:\ v = 0 \text{ on } \partial\Omega \text{ and } v \geq \psi \text{ on } \Omega\}.$$

Thus an alternative problem for u is to solve the constrained optimisation problem

$$E(u) = \min_{v \in K} E(v). \tag{4.8}$$

In the next two sections we will prove the equivalence of (4.8) and (4.1)-(4.6) using the theory of variational inequalities and elliptic partial differential equations, but we first note that (4.1, 4.2, 4.5, 4.6) can be summarised in the formal statement

$$u \geq \psi, \qquad -\nabla^2 u - f \geq 0, \qquad (u-\psi)(\nabla^2 u + f) = 0 \tag{4.9}$$

almost everywhere in Ω. Motivated by the mathematical programming literature, to which we will refer in section 2, we say that (4.9) is an example of a linear complementarity system and that the problem of finding a sufficiently smooth u satisfying (4.9) is a linear complementarity problem.

2. VARIATIONAL INEQUALITIES

It was observed in Chapter II that variational inequalities arise formally as "Euler equations" for inequality constrained problems in the calculus of variations. In general we will see that a minimiser of a convex functional over a convex set of a vector space will indeed satisfy a variational inequality, whereas a minimiser over the whole vector space makes the variational equation of the functional vanish, and hence satisfies the relevant Euler equations.

Before considering an abstract variational inequality on a Hilbert space, let us first give two finite-dimensional examples. We recall that a subset K of a vector space is said to be convex if for every pair $\{\underset{\sim}{x},\underset{\sim}{y}\}$ of vectors belonging to K, $\lambda\underset{\sim}{x} + (1-\lambda)\underset{\sim}{y} \in K$ for all $\lambda \in [0,1]$.

Examples

(i) Let f be a real valued twice continuously differentiable function on $I = [a,b]$ such that $f'' \geq 0$. Then the problem of finding $x^* \in I$ such that

$$f(x^*) = \min_{x \in I} f(x) \tag{4.10}$$

can, by considering the three cases in Fig. 20, be summarised as

$$x^* \in I:\ f'(x^*)(x-x^*) \geq 0 \text{ for all } x \in I. \tag{4.11}$$

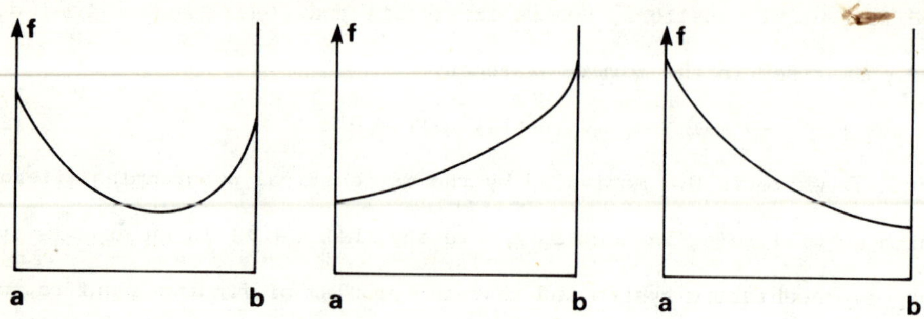

Fig. 20. Problem (4.10).

The inequality (4.11), in which x varies over I, is a simple example of what is called a variational inequality.

(ii) Let $F(\underset{\sim}{x})$ be a real valued twice continuously differentiable convex function defined on a closed bounded convex region K of $\mathbb{R}^N$. A necessary and sufficient condition for F to be convex is that the Hessian matrix of second derivatives $\partial^2 F/\partial x_i \partial x_j$ is positive semi-definite on K. Then consider the optimisation problem

$$F(\underset{\sim}{x}^*) = \min_{\underset{\sim}{x} \in K} F(\underset{\sim}{x}). \tag{4.12}$$

Either $\underset{\sim}{x}^*$ is in the interior of K, in which case $\nabla F(\underset{\sim}{x}^*) = \underset{\sim}{0}$ or $\underset{\sim}{x}^*$ is on the boundary of K, in which case the directional derivative in an admissible direction, i.e. one pointing into the interior of K, must be non-negative. Now K is convex so, for any $\underset{\sim}{x} \in K$, $\underset{\sim}{x}^* + \lambda(\underset{\sim}{x}-\underset{\sim}{x}^*)$, $\lambda \in (0,1]$, defines such a direction and so

$$\nabla F(\underset{\sim}{x}^*)^T(\underset{\sim}{x}-\underset{\sim}{x}^*) \geq 0 \text{ for all } \underset{\sim}{x} \in K\ . \tag{4.13}$$

Conversely, if $\underset{\sim}{x}^*$ satisfies (4.13), then since the Taylor series for $F(\underset{\sim}{x})-F(\underset{\sim}{x}^*)$ begins with

$$\nabla F(\underset{\sim}{x}^*)^T(\underset{\sim}{x}-\underset{\sim}{x}^*) + \tfrac{1}{2}(\underset{\sim}{x}-\underset{\sim}{x}^*)^T\left(\frac{\partial^2 F}{\partial x_i \partial x_j}\right)(\underset{\sim}{x}-\underset{\sim}{x}^*)$$

and the Hessian matrix is positive semi-definite in K, $\underset{\sim}{x}^*$ satisfies (4.12). Thus the optimisation problem and variational inequality are equivalent.

As a particular case let $K \in \mathbb{R}^N_+ = \{\underset{\sim}{x} \in \mathbb{R}^N : \underset{\sim}{x} \geq \underset{\sim}{0}\}$, where we use the convention that $\underset{\sim}{x} \geq \underset{\sim}{0}$ means each component of $\underset{\sim}{x}$ is non-negative. Then the problem of determining $\underset{\sim}{x}^*$ satisfying

$$\nabla F(\underset{\sim}{x}^*) \geq \underset{\sim}{0}, \qquad \underset{\sim}{x}^* \geq \underset{\sim}{0}, \qquad \nabla F(\underset{\sim}{x}^*)^T \underset{\sim}{x}^* = 0 \tag{4.14}$$

is called a nonlinear complementarity problem, [166], the adjective linear only being used when F is quadratic. We can show that (4.13) and (4.14) are equivalent because, if $\underset{\sim}{e}_i$ are coordinate vectors in $\mathbb{R}^N$, taking $\underset{\sim}{x} = \underset{\sim}{x}^* + \underset{\sim}{e}_i$ in (4.13) immediately implies that $\nabla F(\underset{\sim}{x}^*) \geq \underset{\sim}{0}$ and taking $\underset{\sim}{x} = \underset{\sim}{0}, 2\underset{\sim}{x}^*$ yields the complementarity condition $\underset{\sim}{x}^{*T} \nabla F(\underset{\sim}{x}^*) = 0$. Conversely, if $\underset{\sim}{x}^*$ satisfies (4.14), then by complementarity $\nabla F(\underset{\sim}{x}^*)^T (\underset{\sim}{x} - \underset{\sim}{x}^*) = \nabla F(\underset{\sim}{x}^*)^T \underset{\sim}{x}$, which is non-negative for all $\underset{\sim}{x} \geq 0$ since $\nabla F(\underset{\sim}{x}^*) \geq \underset{\sim}{0}$.

When $F(\underset{\sim}{x})$ is the quadratic form $\frac{1}{2}\underset{\sim}{x}^T A \underset{\sim}{x} - \underset{\sim}{b}^T \underset{\sim}{x}$, where A is a symmetric positive semi-definite matrix, (4.13) becomes the variational inequality

$$(A\underset{\sim}{x}^* - \underset{\sim}{b})^T (\underset{\sim}{x} - \underset{\sim}{x}^*) \geq 0 \tag{4.15a}$$

and (4.14) becomes the linear complementarity problem

$$A\underset{\sim}{x}^* - \underset{\sim}{b} \geq \underset{\sim}{0}, \ \underset{\sim}{x}^* \geq \underset{\sim}{0}, \ (A\underset{\sim}{x}^* - \underset{\sim}{b})^T \underset{\sim}{x}^* = 0. \tag{4.15b}$$

This final example is important because, as we shall see, it arises when we discretise the variational inequality formulation of the obstacle problem. For future reference (4.15b) is trivially equivalent to

$$A\underset{\sim}{u} - (\underset{\sim}{b} + A\underset{\sim}{\psi}) \geq \underset{\sim}{0}, \quad \underset{\sim}{u} \geq \underset{\sim}{\psi}, \quad (A\underset{\sim}{u} - (\underset{\sim}{b} + A\underset{\sim}{\psi}))^T (\underset{\sim}{u} - \underset{\sim}{\psi}) = 0 \tag{4.15c}$$

with $\underset{\sim}{x}^* = \underset{\sim}{u} - \underset{\sim}{\psi}$.

Abstract Variational Inequalities

Let V be a real Hilbert space with scalar product $(.,.)$ and associated norm $\|.\|$. Let $a(.,.): V \times V \to \mathbb{R}$ be a continuous bilinear form on V so that $a(u,v)$ is linear in each of the variables (u,v) and there is a constant $\beta > 0$ such that

$$|a(u,v)| \leq \beta\|u\| \|v\| \text{ for all } u,v \in V.$$

Then we say that a(.,.) is coercive if there exists an $\alpha > 0$ such that

$$a(v,v) \geq \alpha\|v\|^2 \text{ for all } v \in V$$

and symmetric if

$$a(u,v) = a(v,u) \text{ for all } u,v \in V.$$

We note that a symmetric coercive bilinear form defines an inner product on V and $[a(.,.)]^{\frac{1}{2}}$ is a norm equivalent to $\|.\|$ in the sense that for all $v \in V$,

$$\alpha^{\frac{1}{2}}\|v\| \leq [a(v,v)]^{\frac{1}{2}} \leq \beta^{\frac{1}{2}}\|v\|.$$

We also let $\ell : V \to \mathbb{R}$ be a continuous linear mapping so that there exists an $M > 0$ such that

$$|\ell(v)| \leq M\|v\| \text{ for all } v \in V.$$

Finally we say that K, a non-empty subset of V, is closed if any convergent sequence in K has its limit in K.

We now consider the variational inequality

(*P1*) Find $u \in K$ such that

$$a(u,v-u) \geq \ell(v-u) \text{ for all } v \in K \tag{4.16}$$

where a(.,.) is a continuous coercive bilinear form, $\ell(.)$ a continuous linear mapping and K a closed convex non-empty subset of V.

If a(.,.) is symmetric, we also consider the optimisation problem

(*P2*) Find $u \in K$ such that $J[u] = \min J[v]$, where $J[v] = \frac{1}{2}a(v,v)-\ell(v)$.

Theorem IV.1, [226]

There exists a unique solution to (*P1*). If a(.,.) is symmetric, then (*P1*) and (*P2*) are equivalent.

Proof

(a) a(.,.) symmetric

We first give a geometric proof of existence and uniqueness based on the projection theorem for Hilbert spaces, which states that for any $w \in V$, there exists a unique element u in a closed convex set K of V which minimises the distance of w from K, (A).

First, since a(.,.) defines an inner product on V, by the Riesz representation theorem, (A), there exists a unique $w_\ell \in V$ such that

$$a(w_\ell,v) = \ell(v) \text{ for all } v \in V ,$$

and so we can write

$$J[v] = \tfrac{1}{2}\, a(w_\ell-v,w_\ell-v) - \tfrac{1}{2}\, a(w_\ell,w_\ell).$$

Hence, (*P2*) is equivalent to minimising the distance, defined by the norm $[a(.,.)]^{\frac{1}{2}}$, from w_ℓ to K. Then, by the Hilbert space projection theorem, there exists a unique solution to (*P2*). The situation is as shown in Fig. 21, where we see that the angle between $w_\ell-u$ and $v-u$ is obtuse, which suggests the variational inequality $a(w_\ell-u,v-u) \leq 0$. Hence, by the definition of w_ℓ, (4.16) is satisfied.

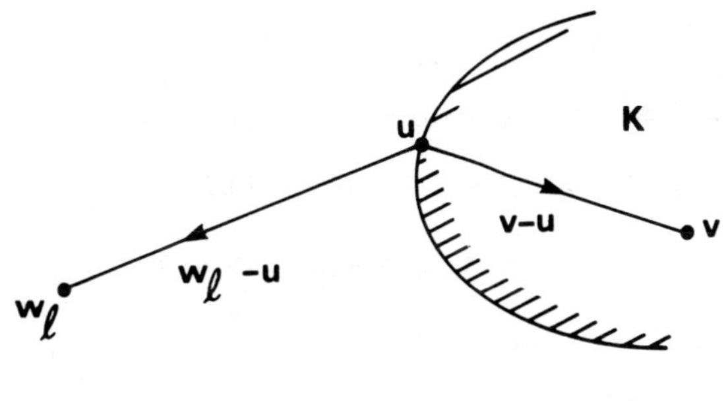

Fig. 21. Projection Theorem

We now prove the equivalence of (*P1*) and (*P2*). First, let u solve (*P1*). Then, for all $v \in K$,

$$J[u] - J[v] = \ell(v-u) - a(u,v-u) - \tfrac{1}{2}\, a(u-v,u-v) \leq 0$$

since $a(u-v,u-v) \geq \alpha\|u-v\|^2$. Hence u solves (*P2*). Next, let u solve (*P2*). Then, since K is convex, v and $(1-\lambda)u + \lambda v$ belong to K for $\lambda \in (0,1]$ and

$$J[u] \leq J[(1-\lambda)u + \lambda v].$$

Expanding for small λ,

$$0 \le \lambda(a(u,v-u) - \ell(v-u)) + \tfrac{1}{2}\lambda^2\ a(u-v,u-v)$$

and so, dividing by λ and letting $\lambda \to 0+$, we retrieve (4.16).

(b) a(.,.) nonsymmetric

Following [121], we now use a contraction mapping argument. Again by the Riesz representation theorem, given $a(.,.)$, $\ell(.)$ and a fixed $u \in V$, there exists a bounded linear operator $A: V \to V$ and an $\tilde{\ell} \in V$ such that

$$(Au,v) = a(u,v) \text{ and } (\tilde{\ell},v) = \ell(v) \text{ for all } v \in V.$$

Thus *(P1)* is equivalent to finding $u \in K$ such that for $\rho > 0$,

$$(u - u + \rho(Au-\tilde{\ell}),\ v-u) \ge 0 \text{ for all } v \in K. \tag{4.17}$$

Now let $P_K(.)$ denote the projection operator from V to K for the inner product $(.,.)$, (A). Writing (4.17) as $(u-x,\ v-u) \ge 0$ and noting that $P_K(x)$ is that element of K which is such that $\|x-P_K(x)\|$ is minimised, (4.17) is equivalent to finding $u \in K$ such that

$$u = P_K\ (u - \rho(Au-\tilde{\ell})) \text{ for all } \rho > 0. \tag{4.18}$$

However, for $\rho < 2\alpha/\|A\|^2 = \rho_0$ say, it can be shown that the mapping in (4.18) is a strict contraction, [121] and so, by the contraction mapping principle, the fixed point problem (4.18) has a unique solution for $\rho < \rho_0$, and it may be obtained by the iteration

$$u^{n+1} = P_K(u^n - \rho(Au^n-\tilde{\ell})).$$

Corollary IV.1

If A is a symmetric positive definite matrix, there exists a unique solution to (4.15).

Proof

When V is the Euclidean space $\mathbb{R}^N$, then defining, for $\underset{\sim}{u}$, $\underset{\sim}{v} \in \mathbb{R}^N$, $a(\underset{\sim}{u},\underset{\sim}{v}) = \underset{\sim}{u}^T A\underset{\sim}{v}$, $\ell(\underset{\sim}{v}) = \underset{\sim}{b}^T\underset{\sim}{v}$ and $K = \{\underset{\sim}{v} \in \mathbb{R}^N\colon \underset{\sim}{v} \ge \underset{\sim}{\psi}\}$, this corollary follows immediately from theorem IV.1.

3. VARIATIONAL INEQUALITY FORMULATION OF THE OBSTACLE PROBLEM

The energy functional $E(v)$, (4.7), involves the integrals of squares of first derivatives. We therefore seek solutions in the space $V \equiv H_0^1(\Omega)$. As usual, let

$$a(u,v) = \int_\Omega \nabla u.\nabla v \, dxdy, \quad \ell(v) = \int_\Omega f\, v \, dxdy = (f,v), \tag{4.19}$$

$$K = \{v \in V: v \geq \psi \text{ almost everywhere in } \Omega\} \tag{4.20}$$

and suppose that

$$f \in L^2(\Omega), \quad \psi \in H^2(\Omega) \cap C^0(\bar{\Omega}) \text{ and } \psi < 0 \text{ on } \partial\Omega. \tag{4.21}$$

Since $a(v,v) = |v|_1^2$ and $|v|_1$ is a norm for $H_0^1(\Omega)$, by the Poincaré inequality (A) the symmetric bilinear form $a(.,.)$ is coercive on $H_0^1(\Omega)$. The continuity of $a(.,.)$ and $\ell(.)$ follow from the Cauchy-Schwarz inequality:

$$\left|\int_\Omega fv \, dx\, dy\right| \leq \left(\int_\Omega |f|^2 \, dxdy\right)^{\frac{1}{2}} \left(\int_\Omega |v|^2 \, dxdy\right)^{\frac{1}{2}} = |f|_0 \, |v|_0 \, ,$$

$$\left|\int_\Omega \nabla u.\nabla v \, dxdy\right| \leq \left(\int_\Omega |\nabla u|^2 \, dxdy\right)^{\frac{1}{2}} \left(\int_\Omega |\nabla v|^2 \, dxdy\right)^{\frac{1}{2}} = |u|_1 \, |v|_1 .$$

Moreover it can be shown that $\psi^+ = \frac{1}{2}(\psi + |\psi|) \in H_0^1(\Omega)$, [140], and since $\psi^+ \geq \psi$, K is non empty. That K is closed follows from observing that if $v_n \to v$ in $H_0^1(\Omega)$, then there exists a subsequence $v_k \to v$ almost everywhere (a.e.) in Ω and if $v_k \geq \psi$ a.e. in Ω, then $v \geq \psi$ a.e. in Ω. Finally, if $v, w \in K$ and $\lambda \in [0,1]$, then

$$\lambda v + (1-\lambda) w \geq \lambda\psi + (1-\lambda)\psi = \psi$$

so K is convex. Thus the conditions of theorem IV.1 are satisfied and we have the following theorem.

Theorem IV.2

There exists a unique $u \in K$ such that

(i) $E(u) \leq E(v)$ for all $v \in K$ (4.22a)

or equivalently

(ii) $\int_\Omega \nabla u.\, \nabla(v-u) \, dxdy \geq \int_\Omega f(v-u) \, dxdy$ for all $v \in K$. (4.22b)

Having thus shown the existence of a unique u which minimises the energy functional over the set of admissible displacements, we will now relate this u to the solution of the free boundary problem (4.1)-(4.6). To do this we first require a certain amount of regularity for u.

Theorem IV.3 [38, 148]

Let Ω have a smooth boundary. If $f \in L^p(\Omega)$ and $\psi \in W^{2,p}(\Omega)$ for any $p \in (1,\infty)$ then the solution u to (4.22) lies in $W^{2,p}(\Omega)$.

Proof

The theorem may be proved by considering a family of unconstrained minimisation problems which penalise the violation of the constraint. This is another example of regularisation or smoothing of the free boundary (p. 30). Let

$$E_\varepsilon[v] = \tfrac{1}{2}\, a(v,v) - (f,v) + \frac{1}{2\varepsilon}\,((v-\psi)^-,\ v-\psi), \tag{4.23}$$

where $(v-\psi)^- = \min(v-\psi, 0)$, and consider

$$u_\varepsilon \in H_0^1(\Omega) : E_\varepsilon(u_\varepsilon) \le E_\varepsilon(v) \text{ for all } v \in H_0^1(\Omega).$$

The variational *equation* for (4.23) is

$$a(u_\varepsilon, v) - (f,v) + \frac{1}{\varepsilon}((u_\varepsilon-\psi)^-,\ v) = 0$$

which is a weak form of the elliptic partial differential equation

$$-\nabla^2 u_\varepsilon + \frac{1}{\varepsilon}\,(u_\varepsilon-\psi)^- = f \text{ in } \Omega \tag{4.24a}$$

with

$$u_\varepsilon = 0 \text{ on } \partial\Omega. \tag{4.24b}$$

Indeed, (4.24) is a special case of the vortex flow problem (1.26), (2.42). Here the "vorticity distribution" $f-1/\varepsilon(u_\varepsilon-\psi)^-$ is such that there is a unique solution for u_ε, because $E_\varepsilon[.]$ is a strictly convex continuous functional on $H_0^1(\Omega)$ and so we can apply theorem III.3 to the minimisation of (4.23). Moreover, it can be shown that $|u_\varepsilon - u|_1 = 0(\varepsilon^{\frac{1}{2}})$ as $\varepsilon \to 0$, [24], so that $u_\varepsilon \to u$ as $\varepsilon \to 0$. Regularity results for elliptic equations, [24], imply that $u_\varepsilon \in W^{2,p}(\Omega)$ and since the estimates of the norm of u_ε in $W^{2,p}(\Omega)$ are independent of ε, the regularity result for u follows.

We are now in a position to prove our principal theorem for the obstacle problem.

Theorem IV.4

(i) Let $f \in L^p(\Omega)$ and $\psi \in W^{2,p}(\Omega)$ for some $p > 2$. Then the solution of (4.22) solves the free boundary problem (4.1)-(4.6) and the linear complementarity problem (4.9).

(ii) Let $f \in L^2(\Omega)$ and $\psi \in H^2(\Omega)$. Suppose $\{u,\Gamma\}$ is the solution of the free boundary problem (4.1)-(4.6) such that $u \in H_0^1(\Omega) \cap H^2(\Omega) \cap C^1(\bar{\Omega})$, and that the closed curves which comprise Γ are smooth. Then u solves the variational inequality (4.22b).

Proof

(i) From the regularity theorem IV.3, $u \in W^{2,p}(\Omega)$ for some $p > 2$, so the Sobolev embedding theorem, (A), implies that $u \in C^1(\bar{\Omega})$. Hence we may define

$$\Omega^+ = \{\underset{\sim}{x}: u(\underset{\sim}{x}) > \psi(\underset{\sim}{x})\}, \quad \Omega^0 = \{\underset{\sim}{x}: u(\underset{\sim}{x}) = \psi(\underset{\sim}{x})\}, \quad \Gamma = \partial\Omega^+ \cap \partial\Omega^0,$$

and it remains to check (4.1) and (4.6).

Let $\phi \in C_0^\infty(\Omega^+)$. Then there is a sufficiently small positive ε such that $u \pm \varepsilon\phi > \psi$ in Ω^+ and $u \pm \varepsilon\phi \geq \psi$ in Ω. Thus, substituting in (4.22b),

$$\int_{\Omega^+} \nabla u . \nabla\phi \, dxdy = \int_{\Omega^+} f\phi \, dxdy \text{ for all } \phi \in C_0^\infty(\Omega^+),$$

which, after integrating by parts, implies that

$$-\nabla^2 u = f \qquad \text{a.e. in } \Omega^+,$$

since $u \in H^2(\Omega)$ and $f \in L^2(\Omega)$.

Similarly, let $\phi \in C_0^\infty(\Omega^0)$ be non-negative so that $u + \phi \geq \psi$ in Ω. Substituting into (4.22b) gives

$$\int_{\Omega^0} \nabla u.\nabla\phi \, dxdy \geq \int_{\Omega^0} f\phi \, dxdy$$

so that $\int_{\Omega^0} (-\nabla^2 u - f)\phi \, dxdy \geq 0$ for all $\phi \geq 0$, $\phi \in C_0^\infty(\Omega^0)$. Thus

$$-\nabla^2 u - f \geq 0 \text{ a.e. in } \Omega^0.$$

The linear complementarity system (4.9) is satisfied since $u \in H^2(\Omega)$, $\psi \in H^2(\Omega)$ and $f \in L^2(\Omega)$.

(ii) For all $v \in K$, we may use the smoothness of Γ and Greens theorem to show that

$$\int_{\Omega} \nabla u.\nabla(v-u)\,dxdy = \int_{\Omega^+ \cup \Omega^0} \nabla u.\nabla(v-u)\,dxdy$$

$$= \int_{\Omega^+} (v-u)(-\nabla^2 u)\,dxdy + \int_{\Omega^0} (v-\psi)(-\nabla^2\psi)\,dxdy + \int_{\partial\Omega^+} \frac{\partial u}{\partial n}(v-u)\,ds + \int_{\partial\Omega^0} \frac{\partial\psi}{\partial n}(v-\psi)\,ds$$

$$= \qquad I_1 \qquad + \qquad I_2 \qquad + \qquad I_3 \qquad + \qquad I_4,$$

say. Now (4.1) gives

$$I_1 = \int_{\Omega^+} f(v-u)\,dxdy$$

and, from (4.6),

$$I_2 \geq \int_{\Omega^0} f(v-u)\,dxdy$$

since $v \geq \psi$ in Ω and $u = \psi$ in Ω^0. Also, on $\partial\Omega^+ \cap \partial\Omega$, $v = u = 0$, and on $\partial\Omega^+ \cap \partial\Omega^0$, $\nabla u = \nabla\psi$ and $u = \psi$. Finally, since $\psi < 0$ on $\partial\Omega$, $\partial\Omega^0$ does not intersect $\partial\Omega$ so that $I_3 + I_4 = 0$. Hence the solution to the free boundary problem satisfies the variational inequality (4.22b).

Concerning theorem IV.3, we now see that we cannot expact more continuity for u than $u \in C^1(\bar{\Omega})$ since $\nabla^2 u$ jumps from $-f$ to $\nabla^2\psi$ across the free boundary

Γ.

The results of theorems IV.2,3,4 may be extended to variable coefficient second order elliptic operators and convex sets of the form

$$K = \{v \in H^1(\Omega): v = g \text{ on } \partial\Omega \text{ and } v \geq \psi \text{ in } \Omega\}.$$

The required smoothness of the fixed boundary $\partial\Omega$ may also be relaxed, and some results are available for the smoothness of Γ. A clear account of the theory for more general elliptic variational inequalities is given in [140].

4. THE DAM PROBLEM

The obstacle problem for an elastic membrane provides a paradigm for other free boundary problems which may be formulated as variational inequalities, perhaps after a preliminary transformation of the dependent variable. An example is the classical problem of seepage through a porous dam, for which an outline of the variational inequality formulation has been given on pp. 58 - 59. Referring to Fig. 9, the free boundary problem is to find the pressure p and free surface Γ such that

$$\Omega^+ = \{\underset{\sim}{x} \in \Omega: p(\underset{\sim}{x}) > 0\}, \text{ the saturated region } y < h(x), \tag{4.25a}$$

$$\Omega^0 = \{\underset{\sim}{x} \in \Omega: p(\underset{\sim}{x}) = 0\}, \text{ the dry region } y > h(x), \tag{4.25b}$$

$$\Gamma = \{\underset{\sim}{x}: y = h(x)\}, \tag{4.25c}$$

$$\nabla^2 p = 0 \text{ in } \Omega^+,$$

$$p = H_1 - y \text{ on } 0A,\ p = H_2 - y \text{ on } CD,\ p = 0 \text{ on } BC,\ \frac{\partial p}{\partial y} = -1 \text{ on } 0D, \tag{4.26}$$

and

$$p = 0 = \frac{\partial}{\partial n}(p + y) \text{ on } \Gamma. \tag{4.27}$$

We have normalised so that $\rho g = 1$ and put $|OA| = H_1$, $|DC| = H_2$.

The "Baiocchi transformation" is

$$U(x,y) = \begin{cases} \int_y^{h(x)} p(x,y')\,dy' & y < h(x) \\ 0 & y > h(x) \end{cases} \tag{4.28}$$

and, as on p. 56, it is a straightforward formal calculation to show that

$$U > 0, \quad \nabla^2 U = 1 \text{ in } \Omega^+ \tag{4.29a}$$

and

$$U = \frac{\partial U}{\partial n} = 0 \text{ on } \Gamma . \tag{4.29b}$$

Thus U satisfies the linear complementarity system

$$-\nabla^2 U + 1 \geq 0, \quad U \geq 0, \quad U(-\nabla^2 U + 1) = 0 \text{ a.e. in } \Omega . \tag{4.30}$$

The boundary conditions on OA', DB' are found by integration to be

$$U = \tfrac{1}{2}\{(H_1 - y)^+\}^2 \text{ on OA}', \qquad U = \tfrac{1}{2}\{(H_2 - y)^+\}^2 \text{ on DB}' \tag{4.31a}$$

and

$$U = 0 \text{ on A'B'}. \tag{4.31b}$$

The boundary condition on OD may be derived by considering the integrated horizontal flow over any x = const. section which is a constant

$$q = -\int_0^{h(x)} \frac{\partial p}{\partial x}(x,y')\,dy'. \tag{4.32}$$

Now by differentiating (4.28),

$$\frac{\partial^2 U}{\partial x^2}(x,0) = -1 + \left(\frac{\partial p}{\partial x}\frac{dh}{dx} - \frac{\partial p}{\partial y}\right)_\Gamma = 0$$

from (4.27) so that, using (4.31a),

$$U(x,0) = -qx + \tfrac{1}{2}H_1^2 = -q(x-L) + \tfrac{1}{2}H_2^2. \tag{4.33}$$

This implies that $q = (H_1^2 - H_2^2)/2L$, and this latter result can be derived physically from observing that the total horizontal force on the liquid in the dam is

$$\int_0^{H_1} p(0,y')\,dy' - \int_0^{H_2} p(L,y')\,dy' = (H_1^2 - H_2^2)/2,$$

which is equal to the total force exerted by the liquid on the porous matrix $= \int_0^L dx' \int_0^{h(x)} - \partial p/\partial x\,(x',y')\,dy' = Lq.$

We may apply the arguments of theorem IV.4 to (4.30) to show that U solves the variational inequality

$$\int_\Omega \nabla U.\nabla(v-U)\,dxdy \geq \int_\Omega -(v-U)\,dxdy \tag{4.34a}$$

$$\text{for all } v \in K = \{H^1(\Omega): v = g(x,y) \text{ on } \partial\Omega,\ v \geq 0 \text{ in } \Omega\}, \tag{4.34b}$$

where g(x,y) are the boundary values of U.

By studying (4.34), it can be shown that h(x) is a strictly decreasing smooth function in (0,L) and that the free boundary problem (4.25)-(4.27) has a unique solution $p = -\partial U/\partial y$, where U solves (4.34), [11, 12, 140]. This justifies our formal derivation of the relationship between (4.25)-(4.27) and (4.34) and we will give the numerical confirmation of this relationship at the end of the next section by comparing the numerical solution of (4.34) with the closed form solution of (4.25)-(4.27) which we mentioned in Chapter II.

5. NUMERICAL ANALYSIS OF ELLIPTIC VARIATIONAL INEQUALITIES

We begin by considering the numerical solution of the free boundary problem (4.1)-(4.6) by finite difference and finite element methods which are naturally suggested by the formulations (4.9) and (4.22b) respectively.

(i) Discretisations

Finite Difference Approximation

For simplicity we assume that Ω may be covered by a square net. A finite difference scheme is obtained by replacing derivatives with central differences in the linear complementarity problem (4.9). For each mesh point (i,j), where (x,y) = (ih, jh), let

$$Z_{i,j} = (4u_{i,j} - u_{i,j+1} - u_{i,j-1} - u_{i-1,j} - u_{i+1,j})/h^2 \tag{4.35}$$

in the interior of Ω. Then the discretisation is

$$Z_{i,j} - f_{i,j} \geq 0,\quad u_{i,j} \geq \psi_{i,j},\quad (Z_{i,j} - f_{i,j})(u_{i,j} - \psi_{i,j}) = 0 \tag{4.36}$$

for all interior points, and

$$u_{i,j} = 0 \text{ on } \partial\Omega, \tag{4.37}$$

where $\{f_{i,j}, \psi_{i,j}\}$ are the values of $\{f,\psi\}$ at the mesh points and $u_{i,j}$ is the finite difference approximation to $u(ih,jh)$. If f is not defined pointwise, we take $f_{i,j} = (\int_S f(\underset{\sim}{x})d\underset{\sim}{x})/h^2$ where S is a square of side h centred at (ih,jh). The algebraic problem is to solve the linear complementarity system

$$A\underset{\sim}{u} - \underset{\sim}{f} \geq \underset{\sim}{0}, \quad \underset{\sim}{u} \geq \underset{\sim}{\psi}, \quad (A\underset{\sim}{u} - \underset{\sim}{f})^T(\underset{\sim}{u} - \underset{\sim}{\psi}) = 0 \tag{4.38}$$

where $\{\underset{\sim}{u}, \underset{\sim}{f}, \underset{\sim}{\psi}\}$ are the vectors of interior nodal values, A is the symmetric positive definite matrix defined by the five-point difference approximation to $-\nabla^2$ on the grid and we have adopted our usual convention about vector inequalities (p. 97). Corollary IV.1 shows that there exists a unique solution to (4.38) which also solves the quadratic programming problem

$$\min_{\underset{\sim}{v} \geq \underset{\sim}{\psi}} \tfrac{1}{2} \underset{\sim}{v}^T A\underset{\sim}{v} - \underset{\sim}{f}^T \underset{\sim}{v} . \tag{4.39}$$

Finite Element Approximation

Let Ω be a convex polygon and V^h be the finite element space approximating $H_0^1(\Omega)$ consisting of continuous piecewise linear functions on the triangulation defined in Chapter III (p. 70). K is approximated by

$$K^h = \{v = \sum_{j \in J} v_j b_j(\underset{\sim}{x}) \in V^h : v_j \geq \psi(\underset{\sim}{x}_j) \text{ for all } j \in J\}. \tag{4.40}$$

For a finite element discretisation, we consider the variational inequality (4.22b) over K^h, i.e.

$$a(u^h, v-u^h) \geq (f, v-u^h) \text{ for all } v \in K^h, \tag{4.41a}$$

which is equivalent to

$$\min_{v \in K^h} \tfrac{1}{2} a(v,v) - (f,v). \tag{4.41b}$$

In matrix terms, this is

$$\min_{\underset{\sim}{v} \geq \underset{\sim}{\psi}} \tfrac{1}{2} \underset{\sim}{v}^T \hat{A} \underset{\sim}{v} - \hat{\underset{\sim}{f}}^T \underset{\sim}{v}, \text{ or } \hat{A}\underset{\sim}{u} - \hat{\underset{\sim}{f}} \geq \underset{\sim}{0}, \quad \underset{\sim}{u} \geq \underset{\sim}{\psi}, \quad (\hat{A}\underset{\sim}{u} - \hat{\underset{\sim}{f}})^T(\underset{\sim}{u} - \underset{\sim}{\psi}) = 0, \tag{4.42a}$$

where the vectors $\{\underset{\sim}{u},\underset{\sim}{v},\underset{\sim}{\psi}\}$ are the nodal values and

$$\hat{A}_{ij} = a(b_i,b_j), \quad (\hat{f})_i = (f,b_i). \tag{4.42b}$$

As noted in Chapter III, $\hat{A}$ is symmetric positive definite, since a(.,.) is a symmetric coercive bilinear form, and thus (4.42) has a unique solution from Corollary IV.1.

(ii) Error Estimates

The convergence of both finite difference and finite element discretisations may be shown in a similar way to those described for the Stefan problem in Chapter III. Briefly, the finite difference formulation may be used to define a so-called "external" approximation to the variational inequality, whereas the finite element formulation, which is a Galerkin method, is an "internal" approximation. Uniform estimates of the $|.|_0$ and $|.|_1$ norms imply the existence of weakly convergent subsequences in the usual way, and these may be shown to converge strongly to the unique solution of the variational inequality, [13, 60, 123].

The first optimal order error estimates for finite element approximations of elliptic variational inequalities were obtained, using different methods in [102, 167]. Here we follow [102], and derive an error estimate for (4.41).

Theorem IV.5

Let $f \in L^2(\Omega)$, and $u, \psi \in H^2(\Omega)$. Then there is a constant c depending on Ω, f and ψ such that

$$|u - u^h|_1 \leq ch.$$

Proof

From (4.41) we have that

$$\begin{aligned} a(u-u^h,u-u^h) &= a(u-u^h,u-u_I) + a(u-u^h,u_I-u^h) \\ &\leq a(u-u^h,u-u_I) + a(u,u_I-u^h) - (f,u_I-u^h), \end{aligned} \tag{4.43}$$

using (4.41) with $v = u_I \in K^h$ as defined in (3.25). Now since $u \in H^2(\Omega)$ and $u_I - u^h \in H_0^1(\Omega)$, Greens theorem implies that

$$a(u,u_I-u^h) - (f,u_I-u^h) = (-\nabla^2 u-f,u_I-u^h)$$

$$= (-\nabla^2 u-f,(u_I-u) + (u-\psi) + (\psi-\psi_I) + (\psi_I-u^h)). \qquad (4.44)$$

Observing that $(-\nabla^2 u-f,u-\psi) = 0$, $\psi_I - u^h \le 0$ and $-\nabla^2 u - f \ge 0$, (4.43) and (4.44) yield the inequality

$$|u-u^h|_1^2 = a(u-u^h,u-u^h)$$

$$\le a(u-u^h,u-u_I) + (-\nabla^2 u-f,(u_I-u) + (\psi-\psi_I))$$

$$\le |u-u^h|_1\ |u-u_I|_1 + |\nabla^2 u+f|_0\ \{|u_I-u|_0 + |\psi-\psi_I|_0\}, \qquad (4.45)$$

using the Cauchy-Schwartz inequality. Thus,

$$\tfrac{1}{2}|u-u^h|_1^2 \le \tfrac{1}{2}|u-u_I|_1^2 + |\nabla^2 u+f|_0\{|u_I-u|_0 + |\psi-\psi_I|_0\}$$

and so, using the interpolation estimates (3.25) of Chapter III, the theorem follows.

The above error estimate also implies that $|u-u^h|_0 \le ch$, since $u-u^h \in H_0^1(\Omega)$ and the Poincaré inequality holds. This is not optimal, however. and the usual device, [57], for obtaining L^2 error estimates for finite element approximations to elliptic equations is not applicable here. However, optimal L^∞ error estimates for the piece-wise linear finite element approximation of the obstacle problem have been obtained [18, 174] in the form

$$\|u-u^h\|_{L^\infty(\Omega)} \le ch^{2-\varepsilon} \text{ for all } \varepsilon > 0. \qquad (4.46)$$

Nonetheless, this estimate does not guarantee that the set $\Omega_+^h = \{x: u^h(x) > \psi\}$ converges to Ω^+. This is easy to see, for example when $\psi = 0$, because u^h in Fig. 22 satisfies (4.46) but is strictly positive everywhere. In practice, the set Ω_+^h is usually close to Ω^+ and does not cover all of Ω. However, a

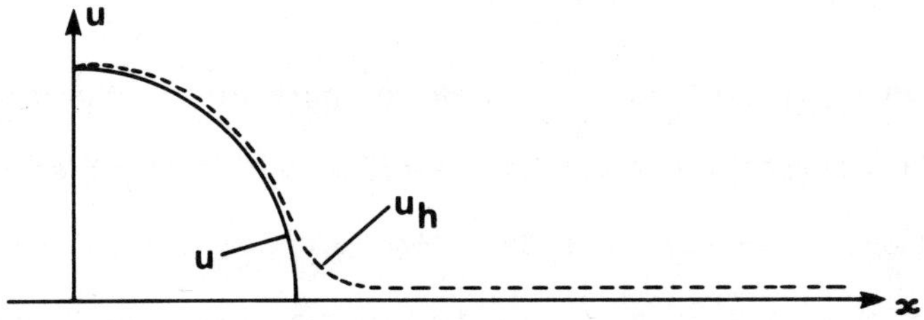

Fig. 22. Functions satisfying (4.46).

result in [17] implies that the set $\Omega_+^h = \{\underset{\sim}{x}: u^h(\underset{\sim}{x}) > \psi + h\}$ does converge to Ω^+ as $h \to 0$ and, for sufficiently small h, $\Omega_+^h \subset \Omega^+$.

Theorem IV.5 can be generalised to variable coefficient elliptic operators with a convex set $K = \{H^1(\Omega): v = g$ on $\partial\Omega$ and $v \geq \psi$ on $\Omega\}$, [41, 102]. Error estimates can also be obtained for polygonal approximations of non-polygonal domains and in cases where numerical integration has to be used to calculate a (.,.) or (f,.), [40, 92, 98]. The accuracy of higher order elements is limited by the regularity of u, which in general has a jump in its second derivatives at the free boundary; for quadratic elements, it is shown in [40, 41] that $\|u-u^h\|_{H^1} \leq ch^{(3/2)-\varepsilon}$.

(iii) <u>Methods for Solving Linear Complementarity Problems (4.38), (4.42)</u>.[62] (4.38) and (4.42) are trivially reducible to the linear complementarity problem

$$\underset{\sim}{u} \geq \underset{\sim}{0}, \quad A\underset{\sim}{u} - \underset{\sim}{b} \geq \underset{\sim}{0}, \quad \underset{\sim}{u}^T(A\underset{\sim}{u}-\underset{\sim}{b}) = 0. \tag{4.47}$$

When A is large, sparse and positive definite, a widely used iterative method for solving (4.47) is the projected successive relaxation method. Given an initial guess $\underset{\sim}{u}^0 \geq \underset{\sim}{0}$, $\underset{\sim}{u}^{k+1}$ is obtained from $\underset{\sim}{u}^k$ by successively calculating, component by component,

$$\bar{u}_i^{k+1} = \{b_i - \sum_{j<i} a_{ij} u_j^{k+1} - \sum_{j>i} a_{ij} u_j^k\}/a_{ii} \tag{4.48a}$$

$$u_i^{k+1} = \max\{0, u_i^k + \omega(\bar{u}_i^{k+1} - u_i^k)\} \tag{4.48b}$$

where ω is a fixed number called the relaxation parameter. The cases $\omega < 1$, $\omega = 1$ and $\omega > 1$ are respectively called the projected underrelaxation, Gauss-Seidel and overrelaxation methods. Together they are called projected SOR methods. The procedure is as simple to program as the well-known SOR method for solving matrix equations arising from elliptic partial differential equations, from which it is generalised. Projected SOR methods were analysed in [78], where the following convergence theorem is proved.

Theorem IV.6

(i) If A is a symmetric positive definite matrix and $\underset{\sim}{u}^0 \geq \underset{\sim}{0}$, then the projected SOR method converges if and only if $\omega \in (0,2)$.

(ii) If strict complementarity holds, i.e.

$$u_i + (A\underset{\sim}{u}-\underset{\sim}{b})_i > 0 \text{ for all } i,$$

then there exists a $k_0(A,\underset{\sim}{b})$ such that $u_i^k > 0$ if $u_i > 0$, and $u_i^k = 0$ if $u_i = 0$ for all $k > k_0$.

The method of proof is to show that (4.48) defines a minimising sequence for the functional $\frac{1}{2}\underset{\sim}{v}^T A\underset{\sim}{v} - \underset{\sim}{b}^T\underset{\sim}{v}$, and then compactness arguments imply convergence. We note that if X denotes the set of indices j such that $u_j = 0$, i.e. X is the set of active constraints, and if

$$A[X] = \{a_{ij}: i \notin X, j \notin X\}, \quad \underset{\sim}{\varepsilon}^k[X] = \{\varepsilon_j^k: j \notin X\}$$

where the error $\underset{\sim}{\varepsilon}^k = \underset{\sim}{u}^k - \underset{\sim}{u}$, then the second part of the theorem implies that

$$\underset{\sim}{\varepsilon}^{k+1}[X] = L_\omega(A[X])\,\underset{\sim}{\varepsilon}^k[X] \text{ for all } k > k_0, \tag{4.49}$$

where $L_\omega(B)$ denotes the standard successive relaxation matrix defined by B and ω. Thus the rate of convergence is asymptotically that of SOR applied to a set of equations with matrix A[X]. It thus seems reasonable, as in Chapter III, p. 78, to define an optimum ω as that which minimises the spectral radius of $L_\omega(A[X])$. Of course X is unknown *a priori* and here we simply suggest two possible choices for ω.

(i) If A is 2-cyclic and consistently ordered, [242], with $a_{ij} \le 0$, $i \ne j$, then $\omega_0(A[X]) \le \omega_0(A)$, where $\omega_0(B)$ denotes the optimum relaxation parameter associated with the matrix B. As it is best to overestimate relaxation parameters, we may reasonably use $\omega = \omega_0(A)$ if X is not too large.

(ii) We can estimate $\omega_0(A[X])$ as the method proceeds, using the current approximation for X at each iteration. The method of [48] for iteratively estimating relaxation factors uses the relationship

$$\omega_0 = 2/(1 + [1 - (p + \omega - 1)^2/p\omega^2]^{\frac{1}{2}}) \tag{4.50}$$

which is valid for 2-cyclic consistently ordered matrices, where p is the largest eigenvalue of L_ω. The idea is to iterate with an $\omega < \omega_0$ for a fixed "block" of iterations, estimate p by $|\underset{\sim}{u}^{k+1} - \underset{\sim}{u}^k|/|\underset{\sim}{u}^k - \underset{\sim}{u}^{k-1}|$ and then update ω using (4.50) until successive estimates agree to some tolerance. This procedure is applicable to (4.47) when (4.49) holds and A is such that (4.50) can be used. Unfortunately, the finite element method does not usually give rise to a 2-cyclic consistently ordered matrix unless the triangulation is regular and linear elements are used.

In Table 3, we present the results of some computations for a matrix satisfying the conditions for (4.50). We observe that an optimum value for ω does exist and that the final value of (4.50) is close to it. The results of either(i) or (ii) are reasonable from the practical viewpoint.

Other variants of point relaxation methods may be found in [83, 158], but here we will only explain in detail how the block SOR method for solving $A\underset{\sim}{u} = \underset{\sim}{b}$ can be modified to give an algorithm for solving (4.47). Let A be partitioned into submatrices A_{ij}, $i,j = 1,\dots,m$, where the diagonal submatrices are square. The vectors $\underset{\sim}{u}^T = (\underset{\sim}{U}_1^T,\dots,\underset{\sim}{U}_m^T)$ and $\underset{\sim}{b}^T = (\underset{\sim}{B}_1^T,\dots,\underset{\sim}{B}_m^T)$ are correspondingly par-

TABLE 3.

Projected SOR for A 2-cyclic and consistently ordered

A of order 465, active set 209			A of order 253, active set 113		
ω	ACT	IT	ω	ACT	IT
1.4	27	88	1.4	24	50
1.5	19	67	1.5	15	36
1.6	11	45	1.56	9	26
1.65	9	35	1.66 (Method (i))	11	34
1.7	13	41	1.7	16	37
1.73 (Method (i))	12	46			
1.8	22	55			
1.9	38	104			
Method (ii)	24	56		24	41
Final ω from (4.50)	1.65			1.56	

ACT = number of iterations to attain active set.

IT = number of iterations to achieve $|u_i^{k+1} - u_i^k| < 10^{-7}$.

---

titioned. Then the modified block SOR method for (4.47) is as follows, [64]. Given an initial guess $\underset{\sim}{u}^0 \geq \underset{\sim}{0}$, $\underset{\sim}{u}^{k+1}$ is obtained from $\underset{\sim}{u}^k$ by successively calculating, block by block, the solution of the complementarity problem

$$A_{ii}\,\bar{\underset{\sim}{U}}_i^{k+1} \geq \underset{\sim}{B}_i - \sum_{j<i} A_{ij}\,\underset{\sim}{U}_j^{k+1} - \sum_{j>i} A_{ij}\,\underset{\sim}{U}_j^{k} = \underset{\sim}{C}_i^{k+1}, \quad \text{say with} \tag{4.51a}$$

$$(A_{ii}\,\bar{\underset{\sim}{U}}_i^{k+1} - \underset{\sim}{C}_i^{k+1})^T\,\bar{\underset{\sim}{U}}_i^{k+1} = 0, \quad \bar{\underset{\sim}{U}}_i^{k+1} \geq \underset{\sim}{0}$$

and the projection

$$\underset{\sim}{U}_i^{k+1} = \underset{\sim}{U}_i^{k} + \omega_i^k(\bar{\underset{\sim}{U}}_i^{k+1} - \underset{\sim}{U}_i^{k}) \tag{4.51b}$$

where $\omega_i^k = \max\{\bar{\omega}\colon \bar{\omega} \leq \omega,\ \underset{\sim}{U}_i^k + \bar{\omega}(\bar{\underset{\sim}{U}}_i^{k+1} - \underset{\sim}{U}_i^k) \geq \underset{\sim}{0}\}$.

The following theorem is proved in [64].

Theorem IV.7

If A is symmetric and positive definite and $\omega \in (0,2)$, then the method (4.51) converges.

At each iteration (4.51a) has to be solved so that this method is likely to be inefficient unless a fast method for solving this subproblem is available. Now matrices arising from discretisations of second order elliptic operators can frequently be partitioned so that the A_{ii} are non-singular tri-diagonal matrices with non-positive off-diagonal elements such that A_{ii}^{-1} has non-negative elements. For such a class of matrices, efficient methods which use principal pivoting and matrix factorisation have been devised, [65, 212]. Frequently the free boundary problem is such that the mesh points may be ordered so that all the non-zero elements of a particular block come first. For example, in the case of a rectangular dam problem, a suitable partition would have the blocks corresponding to vertical lines and the ordering within each block would start from the base of the dam. In these instances, we can give a fast direct method as follows.

Let $\underset{\sim}{q} \in \mathbb{R}^n$, and suppose $M \in \mathbb{R}^{n\times n}$ is the tridiagonal matrix with diagonal elements $a_i > 0$, $i = 1,\dots,n$, super-diagonal elements $c_i < 0$, $i = 1,\dots,n=1$, and sub-diagonal elements $b_i < 0$, $i = 2,\dots,n$, which satisfy

$$a_1 > -c_1; \quad a_n > -b_n; \quad a_j \geq -(b_j + c_j), \quad j = 2,\dots,n-1. \tag{4.52}$$

Consider the complementarity problem

$$\underset{\sim}{w} = M\underset{\sim}{z} + \underset{\sim}{q} \geq \underset{\sim}{0}, \quad \underset{\sim}{z} \geq \underset{\sim}{0}, \quad \underset{\sim}{w}^T\underset{\sim}{z} = 0. \tag{4.53}$$

Proposition

(i) There exists a unique solution to (4.53).

(ii) If $q_i < 0$, $i = 1,2,\dots,k_0$, and $q_i \geq 0$, $i = k_0+1,\dots,n$, then there exists a $k > k_0$ such that $z_i = 0$, $i > k$ and $z_i > 0$, $i \leq k$.

Proof

(i) M is an irreducible Minkowski matrix, and so the result follows from a theorem in [65].

(ii) If $q_i < 0$, then $z_i \geq -(q_i + c_i z_{i+1} + b_i z_{i-1})/a_i > 0$, since a_i, $-b_i$ and $-c_i$ are positive. Thus $z_i > 0$ for $i \leq k_0$. Letting $(k+1)$ be the first integer greater than k_0 at which $z_k = 0$, (4.53) is satisfied by $\{z_i = 0; \omega_i = q_i; i > k\}$.

Suppose now that the solution to (4.53) does indeed satisfy $z_i > 0$, $i \leq k$; $z_i = 0$, $i > k$, so that $\underset{\sim}{z}$ may be determined by solving

$$\left.\begin{aligned} &a_1 z_1 + c_1 z_2 + q_1 = 0 \\ &b_i z_{i-1} + a_i z_i + c_i z_{i+1} + q_i = 0, \; i = 2,\ldots,k-1 \\ &b_k z_{k-1} + a_k z_k + q_k = 0 \end{aligned}\right\} \tag{4.54a}$$

$$z_i > 0, \; i \leq k; \; z_i = 0, \; i \geq k+1; \; b_{k+1} z_k + q_{k+1} \geq 0. \tag{4.54b}$$

We observe that in forward Gaussian elimination for (4.54a), the first component to be found is z_k; thus if z_k also satisfies $b_{k+1} z_k + q_{k+1} \geq 0$, then indeed k has been found and it remains only to backsolve to determine the unique solution to (4.53). Therefore, in algorithmic form the method is:

<u>DECOMPOSE</u> $i = 1, \quad \alpha_1 = a_1, \quad f_1 = -q_1/a_1$

$$\left.\begin{aligned} \text{STEP:}\quad & i \to i+1 \\ & \gamma_{i-1} = c_{i-1}/\alpha_{i-1} \\ & \alpha_i = a_i - b_i \gamma_{i-1} \\ & f_i = -(q_i + b_i f_{i-1})/\alpha_i \end{aligned}\right\} \tag{4.55a}$$

IF $f_i > -q_{i+1}/b_{i+1}$ THEN GO TO STEP

$$\left.\begin{aligned} \underline{\text{BACKSOLVE}}\quad & k = i \\ & z_k = f_k \\ & i = k-1 \; (-1) \; 1 \end{aligned}\right\} \tag{4.55b}$$

$$z_i = f_i - \gamma_i z_{i+1}.$$

This procedure is numerically stable since M is diagonally dominant [131 , p. 55]. The operation count is (5k+4) + (k−1) multiplications and divisions, the latter term coming from testing the inequality for f_i. This compares with the counts $k^2 + 3k$ and $(5k-4) + \frac{3}{2}(k^2-k_0^2) + \frac{3}{2}(k-k_0)$ for the methods of [65 , 212] applied to this problem. In the applications, $k-k_0 = O(1/h)$ and generally $k \gg k_0$. Also, since $k = O(1/h)$, halving h doubles the operation count for (4.55).

6. NUMERICAL EXAMPLES

We now present four examples of the application of finite difference and finite element discretisations to free boundary problems. The first two are model problems with simple closed form solutions.

These first two examples concern problems with non-zero Dirichlet data on $\partial\Omega$ but, for ease of presentation, all the preceding analysis of elliptic variational inequalities has assumed homogeneous fixed boundary data. To apply our theory to problems with non-zero Dirichlet data g on $\partial\Omega$, all that is needed is the existence of $G(x) \in H^2(\Omega)$ such that $G = g$ on $\partial\Omega$. Then the variational inequality for

$$u \in \tilde{K} = \{H^1: \tilde{v} = g \text{ on } \partial\Omega,\ \tilde{v} \geq \psi \text{ a.e. in } \Omega\}$$

which is such that

$$\int_\Omega \nabla u.\ \nabla(\tilde{v}-u)\,dx \geq \int_\Omega f(\tilde{v}-u)\,d\underset{\sim}{x} \tag{4.56}$$

for all $\tilde{v} \in \tilde{K}$ can be written as one for

$$w \in K = \{H_0^1(\Omega): v \geq \psi - G \text{ a.e. in } \Omega\}$$

which is such that

$$\int_\Omega \nabla w.\ \nabla(v-w)\,d\underset{\sim}{x} \geq \int_\Omega (f+\nabla^2 G)(v-w)\,d\underset{\sim}{x} \tag{4.57}$$

for all $v \in K$, where $w = u-G$.

1. A one-dimensional model problem

Consider the free boundary problem for $\{u(x), x_0\}$ such that

$$\frac{d^2u}{dx^2} = 1, \quad 0 < x < x_0 \tag{4.58}$$

with $u(0) = \alpha$, $0 < \alpha < \frac{1}{2}$, $u(x_0) = \frac{du}{dx}(x_0) = 0$. The corresponding linear complementarity problem is

$$-\frac{d^2u}{dx^2} + 1 \geq 0, \; u \geq 0, \; u\left(-\frac{d^2u}{dx^2} + 1\right) = 0 \tag{4.59}$$

for almost all $x \in (0,1)$, with $u(0) = \alpha$, $u(1) = 0$, and the equivalent variational inequality is to find u such that

$$\int_0^1 \frac{du}{dx}\left(\frac{dv}{dx} - \frac{du}{dx}\right) dx \geq \int_0^1 -(v-u)\, dx \tag{4.60}$$

for all $v \in K \equiv \{H^1(0,1);\ v(0) = \alpha,\ v(1) = 0,\ v \geq 0\}$. (4.60) means (4.59) is identical to the constrained minimisation problem

$$\min_{v \in K} J[v] = \left\{\tfrac{1}{2}\int_0^1 \left(\frac{dv}{dx}\right)^2 dx + \int_0^1 v\, dx\right\}, \tag{4.61}$$

and the solution to all these problems is

$$u(x) = \begin{cases} \tfrac{1}{2}(x-x_0)^2 & 0 < x < x_0 \\ 0 & x_0 < x < 1 \end{cases} \tag{4.62}$$

where $x_0 = (2\alpha)^{\frac{1}{2}}$.

The finite difference approximation (p. 107) to (4.59) is

$$(2u_i - u_{i-1} - u_{i+1})/h^2 + 1 \geq 0, \quad u_i \geq 0, \; (2u_i - u_{i-1} - u_{i+1} + h^2)u_i = 0, \tag{4.63}$$

$i = 1, \ldots, N-1$,

with $u_0 = \alpha$, $u_N = 0$, $h = 1/N$,

where u_i is the approximation to $u(ih)$. In matrix form this is

$$A\underset{\sim}{u} = \underset{\sim}{b} + \underset{\sim}{z}, \quad \underset{\sim}{u}^T\underset{\sim}{z} = 0, \quad \underset{\sim}{u} \geq \underset{\sim}{0}, \quad \underset{\sim}{z} \geq \underset{\sim}{0} \tag{4.64}$$

where

$$A = \begin{pmatrix} 2 & -1 & & & & \\ -1 & 2 & -1 & & 0 & \\ & \cdot & \cdot & \cdot & & \\ & & \cdot & \cdot & \cdot & \\ 0 & & & -1 & 2 & -1 \\ & & & & -1 & 2 \end{pmatrix} , \quad \underset{\sim}{b} = \begin{pmatrix} \alpha-h^2 \\ -h^2 \\ \cdot \\ \cdot \\ -h^2 \end{pmatrix} .$$

For the finite element approximation, let u^h solve $\min_{v\in K^h} J[v]$

$$V^h = \{v \in C[0,1]; \quad v_0 = \alpha, \quad v_N = 0; \quad v = \sum_{i=0}^{N} v_i\phi_i\}, \quad K^h = \{v \in V^h : v_i \geq 0\},$$

where ϕ_i, $i = 0,1,\ldots,N$ are piecewise linear basis functions on the uniform partition $(ih, (i+1)h)$ of $[0,1]$ and are defined by

$$\phi_i(x) = \begin{cases} (x-(i-1)h)/h, & x \in [(i-1)h, ih] \\ ((i+1)h-x)/h, & x \in [ih,(i+1)h)] \end{cases} \quad . \tag{4.65}$$

Then, for $v \in V^h$,

$$J[v] = \sum_{i=1}^{N} \int_{(i-1)h}^{ih} \{\tfrac{1}{2}\left(\frac{v_i - v_{i-1}}{h}\right)^2 + v_{i-1} + \frac{(x-(i-1)h)(v_i - v_{i-1})}{h}\} \, dx$$

$$= h^{-1}(\tfrac{1}{2}\underset{\sim}{v}^T A \underset{\sim}{v} - \underset{\sim}{b}^T \underset{\sim}{v}) + \text{const.},$$

where the constant is independent of $\underset{\sim}{v}$, so that the finite difference and finite element methods are identical in this case.

Now let ℓ be the first mesh point at which the discrete solution vanishes. The discrete free boundary problem for $\{u_i, \ell\}$ is

$$u_{i-1} - 2u_i + u_{i+1} = h^2, \quad 1 \leq i < \ell$$

$$u_0 = \alpha, \quad u_\ell = 0, \quad 0 < u_{\ell-1} \leq h^2,$$

which we may solve to find the error

$$u(ih) - u_i = i(x_0 - \ell h)^2/2\ell, \quad i = 0,1,\ldots,\ell \ .$$

The maximum error occurs at the free boundary and the condition on $u_{\ell-1}$ implies that

$$\ell(\ell+1) \ge (x_0/h)^2 > \ell(\ell-1)$$

i.e.

$$|x_0-\ell h| < \begin{cases} h/2, & x_0 > \ell h \\ \ell h/(2\ell-1), & x_0 \le \ell h \end{cases}$$

The error in the free boundary position is thus nearly the optimal value h/2, and $|u_i-u(ih)| = O(h^2)$. The approximation to x_0 may be improved either by linear extrapolation to zero of the last two positive mesh points or quadratic extrapolation using both free boundary conditions. These yield extrapolated free boundary points

$$x_L = (\ell-1)h + u_{\ell-1}h/(u_{\ell-2}-u_{\ell-1}),\quad x_Q = (\ell-1)h + u^{\frac{1}{2}}_{\ell-1}h/(u^{\frac{1}{2}}_{\ell-2}-u^{\frac{1}{2}}_{\ell-1}) \tag{4.66a,b}$$

respectively and Table 4 shows that the latter is much more accurate than the former. This is also true in two dimensions as we shall see in the next example.

TABLE 4.

Extrapolated Free Boundary Position

x_0	N	h	ℓh	x_L	x_Q
0.5	3	0.3333	0.6667	0.3529	0.4400
	5	0.2000	0.6000	0.4080	0.4488
	7	0.1429	0.5714	0.4329	0.4586
	9	0.1111	0.5555	0.4471	0.4647
	15	0.0667	0.5333	0.4677	0.4760
	25	0.0400	0.5200	0.4804	0.4843
0.3333	5	0.2000	0.4000	0.2326	0.3196
	8	0.1250	0.3750	0.2695	0.3227
	10	0.1000	0.3000	0.2460	0.3280
	25	0.0400	0.3200	0.2983	0.3309
	100	0.0100	0.3300	0.3245	0.3327

This approach to the error analysis of one-dimensional problems has been applied to the journal bearing problem of lubrication theory, [79].

2. A two-dimensional model problem

A model obstacle problem is to find a non-negative $u(x,y)$ in $x > 0$, $|y| < 1$ which satisfies

$$\nabla^2 u = 1 \text{ in } \Omega^+, \text{ where } u > 0, \tag{4.67a}$$

with

$$u = \frac{1}{4}(y^2-1) - \frac{1}{2}\log\left(\frac{y^2+1}{2}\right) \text{ on } |y| \le 1, \quad x = 0, \tag{4.67b}$$

and

$$u = \frac{\partial u}{\partial n} = 0 \text{ on the free boundary } \Gamma. \text{ (Fig. 23)} \tag{4.67c}$$

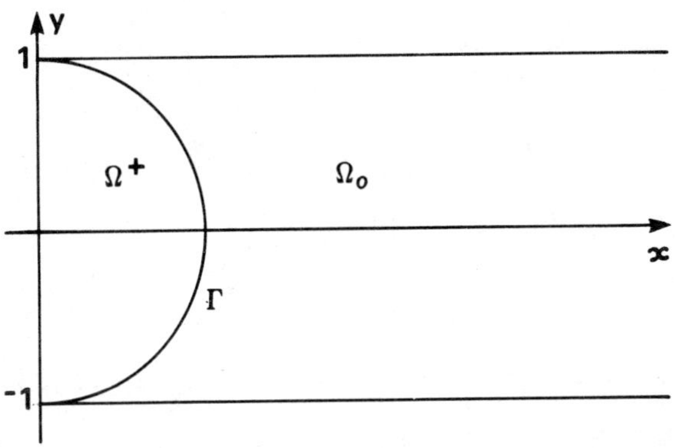

Fig. 23. Obstacle Problem

The solution is

$$u = \frac{(x+1)^2+y^2}{4} - \frac{1}{2} - \frac{1}{2}\log\frac{(x+1)^2+y^2}{2} \text{ in } \Omega^+, \tag{4.68a}$$

$$\text{and on } \Gamma, \ (x+1)^2+y^2 = 2. \tag{4.68b}$$

Taking Ω to be the rectangle $(0,L) \times (-1,1)$, where L is large enough that $\Omega^+ \subset \Omega$, theorem IV.4 and (4.56, 4.57) show that u solves

$$\min_{v\in K} \frac{1}{2}\int_\Omega |\nabla v|^2 \, dxdy + \int_\Omega v \, dxdy$$

where $K = \{v \in H^1(\Omega) : v \geq 0, v \text{ satisfies } (4.67b)\}$.

For this model problem, the minimisation was carried out on the rectangle $(0,\frac{1}{2}) \times (0,1)$, using boundary values from the exact solution (4.68a), and using linear finite elements on a regular triangular mesh, i.e. one with equal mesh lengths in the x and y directions and with triangle sides parallel to x = 0, y = 0, x = y. The accuracy is shown in Table 5, where ε_{max} is the maximum error at mesh points and $\varepsilon_{L_2} = \{\sum_{i,j} (u(ih,jh) - u_{ij})^2 h^2\}^{\frac{1}{2}}$, in the usual notation. Table 5 also confirms the estimate (4.46).

TABLE 5.

Accuracy of finite element solution

h	$\varepsilon_{L_2} \times 10^5$	$\varepsilon_{max} \times 10^4$
1/16	4.20	2.91
1/24	1.84	1.27
1/32	0.85	0.58
1/48	0.22	0.23

The discrete free boundary has the stepped shape shown in Fig. 24, and extrapolation is needed to find Γ more accurately. On an x = const line, taking Γ at the first ordinate at which u vanishes would be accurate at best to within h/2, and this is demonstrated in the first column of Table 6. We may now use linear or quadratic extrapolation in the x or y directions. It is probably best to choose whichever direction is closer to the normal to Γ, but in Table 6 we just present the results of extrapolating in the y direction for small x. Linear extrapolation gives worse results than taking the ordinate of the first zero mesh point, and quadratic extrapolation gives good agreement with the true solution.

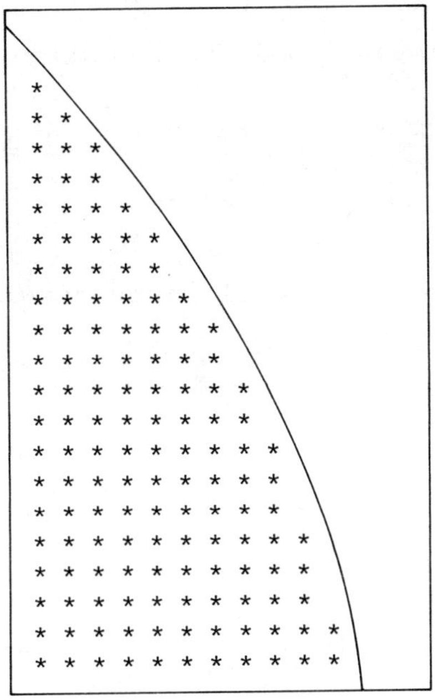

Fig. 24. Discrete Free Boundary h = 1/32.

TABLE 6.

Calculation of Γ, $h = 1/48$. *All values* $\times 10^4$.

x = i/48	First zero ordinate	Linear extrapolation	Quadratic extrapolation	Exact solution
1	9792	9672	9835	9841
3	9583	9407	9494	9508
5	9167	9022	9153	9152
7	8750	8621	8772	8771
9	8333	8207	8362	8360
11	7917	7778	7918	7916
13	7500	7316	7392	7433
15	6875	6747	6899	6902

3. Elastic-plastic torsion of a cylinder

In Chapters I and II (p. 12 and p. 53), we suggested that the determination of the displacement field is equivalent to minimizing

$$J[v] = \tfrac{1}{2} \int_\Omega |\nabla v|^2 dxdy - \int_\Omega 2\theta v \, dxdy \tag{4.69}$$

over $v \in K = \{H_0^1(\Omega) : |\nabla v| \le 1\}$,

where we have set $k = G = 1$ in (1.12, 1.13) for simplicity. The equivalence of this minimisation, the free boundary problem (1.12, 1.13) and the obstacle problem

$$\min_{v \in \hat{K}} J[v]$$

where

$$\hat{K} = \{H_0^1(\Omega) : v \le d(x,y)\} \tag{4.70}$$

and d is the "roof function" for Ω, i.e. the distance from (x,y) to $\partial\Omega$, is proved in [37 , 56]..

Now let Ω be the pentagon with vertices $(\pm\frac{3}{4},0)$, $(\pm 1,\frac{3}{4})$ and (0,1). A finite element calculation can be made on the triangulation of half the pentagon shown in Fig. 25 and the results of a typical computation for $\theta = 3.75$, using the algorithm (4.48), are shown in Fig. 26. The plastic region is denoted by asterisks, and the elastic region by dots.

Further details of the numerical analysis of (4.69) are given in [51, 123].

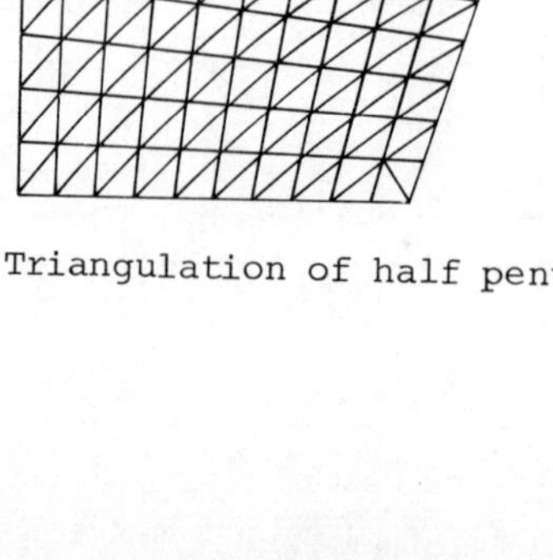

Fig. 25. Triangulation of half pentagon.

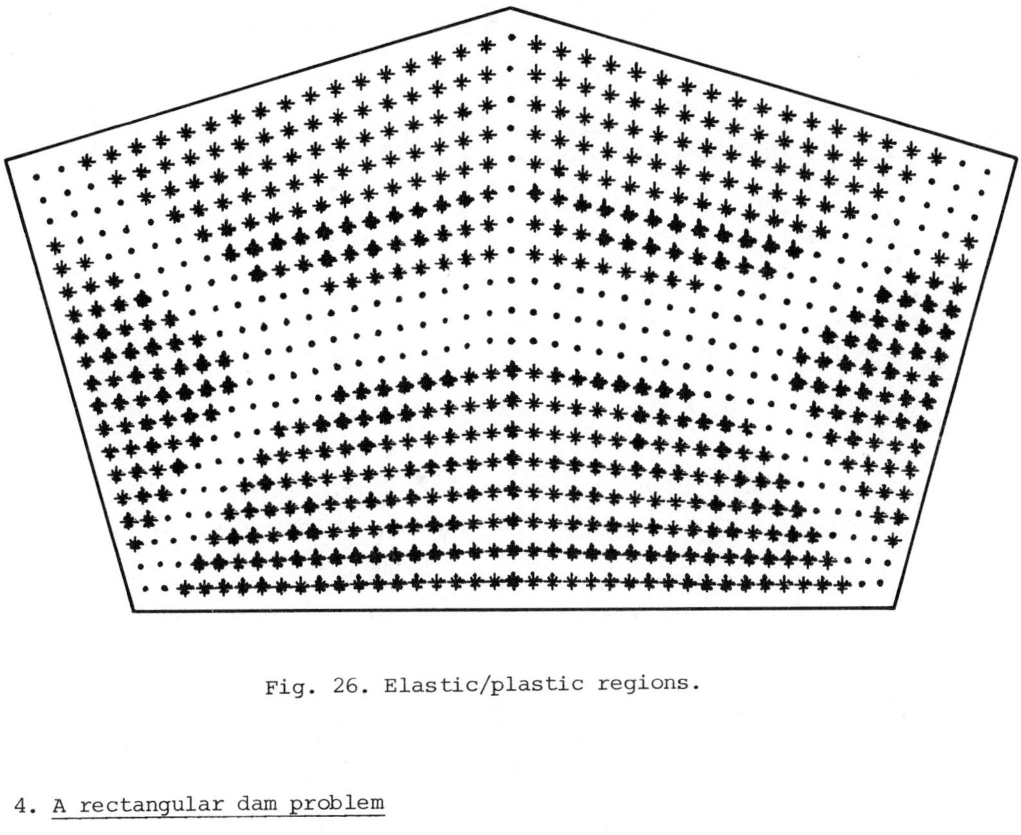

Fig. 26. Elastic/plastic regions.

4. A rectangular dam problem

The inequality (4.34) can be solved using linear finite elements as described for example 2. In the case $H_1 = 1$, $H_2 = 1/6$, $L = 2/3$ the positive and zero mesh points are displayed in Fig. 27. . Again more accurate ordinates for the free surface than those obtainable from Fig. 27 can be found by quadratic extrapolation in the y direction. The result of using (4.66b) when $u_{\ell-1} > u_{\ell-2}/10$, and replacing ℓ by $\ell-1$ otherwise, is given in the final Table. For ease of reference, we also include the analytical solution, [193], of two

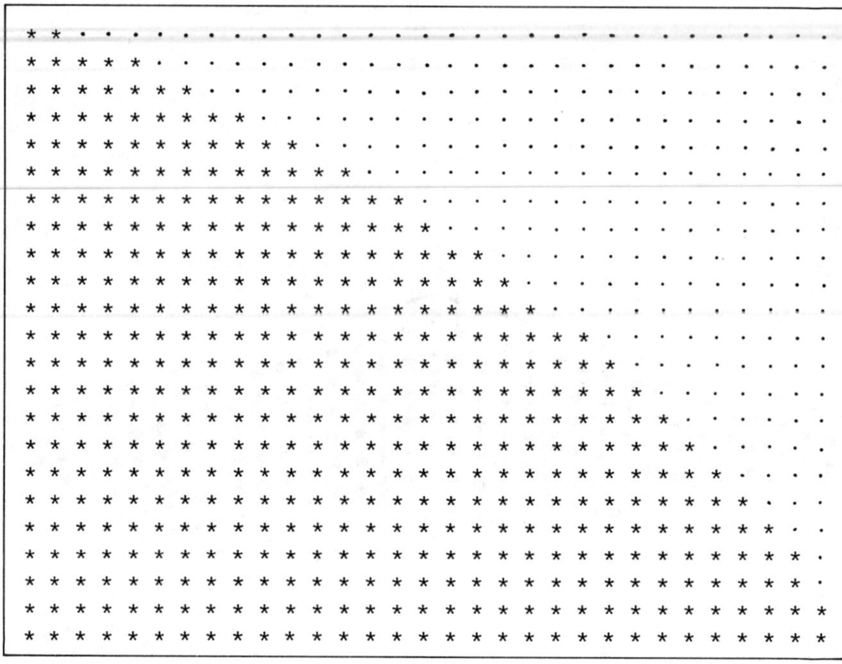

Fig. 27. Rectangular dam problem.

TABLE 7.

Ordinates of free surface for rectangular dam problems

mesh size / i=24x	Numerical Solutions 1/12	1/24	1/48	Analytical Solutions $H_1 = 1$, $H_2 = 1/6$, $L = 2/3$	$H_1 = 1$, $H_2 = 0$, $L = 2/3$	x = i/0.010125	$H_1 = 0.322$, $H_2 = 0.084$, $L = 0.162$
0				1.000000	1.000000	0	0.322000
1		9869	9885	0.989397	0.989347	1	0.319396
2	9704	9735	9743	0.975455	0.975316	2	0.315958
3		9579	9586	0.959326	0.959067	3	0.311971
4	9380	9378	9406	0.941295	0.940881	4	0.307504
5		9202	9207	0.921472	0.920869	5	0.302581
6	8965	8966	8992	0.899884	0.899056	6	0.297210
7		8747	8757	0.876507	0.875413	7	0.291381
8	8486	8496	8505	0.851267	0.849864	8	0.285076
9		8224	8235	0.824042	0.822282	9	0.278264
10	7938	7934	7936	0.794648	0.792480	10	0.270896
11		7605	7607	0.762815	0.760182	11	0.262907
12	7249	7256	7281	0.728140	0.724978	12	0.254195
13		6879	6900	0.689986	0.686226	13	0.244604
14	6433	6467	6475	0.647242	0.642803	14	0.233859
15		6016	5932	0.597518	0.592311	15	0.221378
16				0.529413*	0.523335*	16	0.204385*

*(Separation point)

other dam problems.

7. TIME DEPENDENT PROBLEMS

Although, as mentioned in Chapter II, there is no physically obvious unsteady generalisation of the obstacle problem, we will now consider two moving boundary problems which can be regarded mathematically as "obstacle problems of parabolic type". We will subsequently define and discuss a class of relevant parabolic variational inequalities and conclude with a brief analysis of the Hele-Shaw problem (p. 22 and p. 55).

(i) *The Oxygen Consumption Problem*

Let Ω be an open bounded region in $\mathbb{R}^N$ with a smooth boundary $\partial\Omega$ and suppose that oxygen with concentration $c(\underset{\sim}{x},t)$ is diffusing in Ω and being absorbed volumetrically at a constant rate R (p. 17). This is a moving boundary problem for $\Gamma(t)$ where we define the sets

$$\Omega^+(t) = \{\underset{\sim}{x} \in \Omega: c(\underset{\sim}{x},t) > 0\}, \qquad \Omega^0(t) = \{\underset{\sim}{x} \in \Omega: c(\underset{\sim}{x},t) = 0\} \tag{4.71a}$$

$$\Gamma(t) = \partial\Omega^+(t) \cap \partial\Omega^0(t), \tag{4.71b}$$

$$\Omega_T = \Omega \times (0,T), \quad \Omega_T^+ = \bigcup_{t<T} \Omega^+(t), \quad \Omega_T^0 = \bigcup_{t<T} \Omega^0(t), \quad \Gamma = \bigcup_{t<T} \Gamma(t). \tag{4.71c}$$

Three possible configurations are shown in Fig. 28, depending on whether fixed boundaries (shaded) are present or not.

When $c > 0$, the diffusion equation

$$\frac{\partial c}{\partial t} = \nabla^2 c - R \tag{4.72}$$

holds in Ω_T^+, where we have assumed unit diffusivity, and, as in (1.22), mass conservation at the moving boundary implies that

$$\nabla c = \underset{\sim}{0}, \quad \underset{\sim}{x} \in \Gamma(t). \tag{4.73}$$

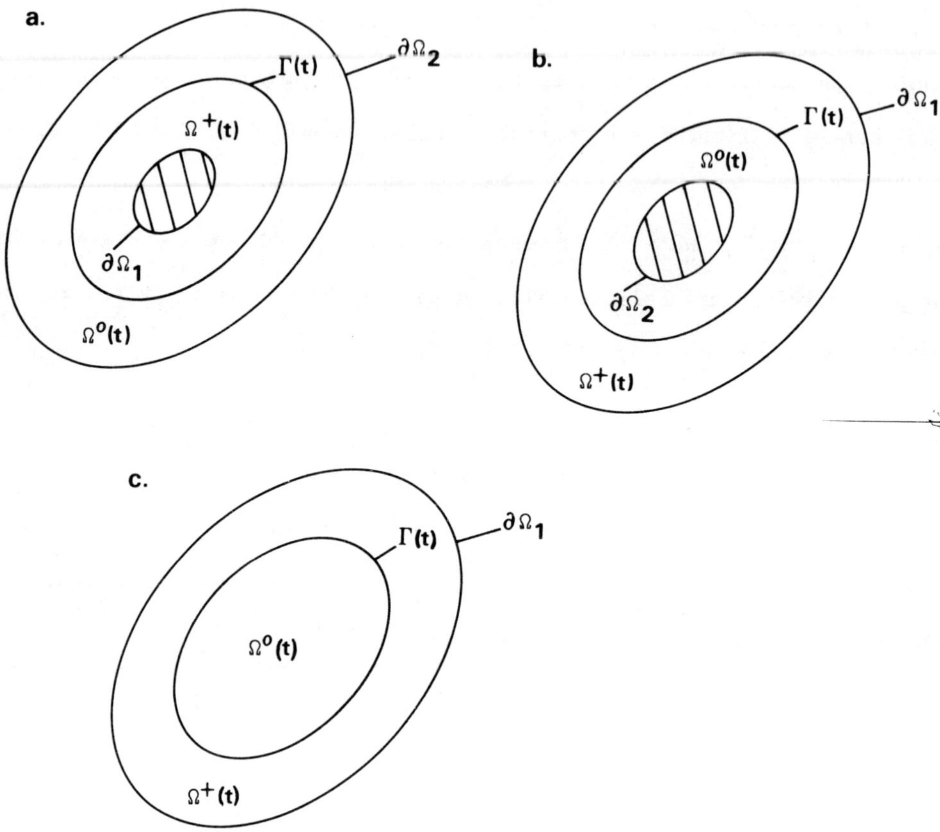

Fig. 28. Oxygen Consumption Problem.

The initial concentration

$$c(\underset{\sim}{x},0) = c_0(\underset{\sim}{x}) \geq 0 \qquad (4.74a)$$

is prescribed in Ω and we suppose for convenience that

$$c = g(\underset{\sim}{x},t) \geq 0, \quad \underset{\sim}{x} \in \partial\Omega, \qquad (4.74b)$$

and that if $\partial\Omega_2$ is non-empty

$$c = 0, \quad \underset{\sim}{x} \in \partial\Omega_2. \qquad (4.74c)$$

If the distributional derivatives $\partial c/\partial t$ and $\nabla^2 c$ exist in $L^2(\Omega_T)$ (A) then, since for all non-negative $\phi \in C_0^\infty(\Omega_T)$

$$\int_{\Omega_T} (\frac{\partial c}{\partial t} - \nabla^2 c + R)\phi d\underset{\sim}{x}dt$$

$$\equiv \int_{\Omega_T} \{\nabla c.\nabla\phi + (\frac{\partial c}{\partial t} + R)\phi\} d\underset{\sim}{x}dt$$

$$= \int_{\Omega_T^0} R\phi\, d\underset{\sim}{x}dt + \int_0^T \int_{\partial\Omega^+(t)} \frac{\partial c}{\partial n} \phi dS_{\underset{\sim}{x}}\, dt \geq 0, \tag{4.75}$$

c satisfies the linear complementarity system

$$\frac{\partial c}{\partial t} - \nabla^2 c + R \geq 0, \quad c \geq 0, \quad (\frac{\partial c}{\partial t} - \nabla^2 c + R)c = 0 \tag{4.76}$$

a.e. in Ω_T.

(ii) *The One-Phase Stefan Solidification Problem*

If a stationary liquid is initially at its freezing temperature and heat is then conducted out of the liquid, the resulting solidification process may be formulated as a one-phase Stefan problem (p. 15). As usual, we assume the temperature $u(\underset{\sim}{x},t)$ is zero at the freezing point and define sets as in (3.1), replacing Ω^+ by Ω^0. Thus $\Omega^-(t)$ is the solid phase and $\Omega^0(t)$ the liquid, and these sets may also be as in Fig. 28. We note that Fig. 28 may represent a problem in an infinite region outside $\partial\Omega_1$ as long as $\partial\Omega_2$ is is chosen to be sufficiently remote to ensure that, in $(0,T)$, $\Omega^+(t) \subset \Omega$.

As on p. 55, let there exist a function $\ell(\underset{\sim}{x}) \geq 0$ on Ω such that

$$\ell(\underset{\sim}{x}) = 0, \quad \underset{\sim}{x} \in \Omega^-(0) \cup \Gamma(0) \tag{4.77a}$$

$$\text{with } \nabla\ell \neq \underset{\sim}{0}, \quad \ell > 0 \text{ and } \ell \text{ smooth in } \Omega^0(0) \tag{4.77b}$$

and with the property that, for $t > 0$,

$$\Omega^-(t) = \{\underset{\sim}{x}:\ell(\underset{\sim}{x}) < t\}, \quad \Omega^0(t) = \{\underset{\sim}{x}:\ell(\underset{\sim}{x}) > t\}, \quad \Gamma(t) = \{\underset{\sim}{x}:\ell(\underset{\sim}{x}) = t\}. \tag{4.77c}$$

This is similar to the assumption in the dam problem (p. 58) that the free boundary meets any vertical line precisely once. Our final assumption is that $\Omega^-(0)$ is connected and attached to $\partial\Omega_1$ and that $\Omega^0(0)$ is connected. Then (4.77) imply that each of $\Omega^-(t)$ and $\Omega^0(t)$ is connected and that, if $0 < t_1 < t_2$

$$\Omega^-(t_1) \subset \Omega^-(t_2) \ . \tag{4.78}$$

This simply states that the frozen region expands when cooling takes place and, with suitable assumptions about the data, the weak solution defined by (3.9) can be proved to satisfy (4.78), [114].

When the thermal parameters are constant, so that the enthalpy is given by (3.18) with $u_F = 0$ and $\alpha_1 = \alpha$, say, the heat conduction equation in the solid is

$$\alpha \frac{\partial u}{\partial t} = \nabla^2 u, \quad \underset{\sim}{x} \in \Omega_T^- \ . \tag{4.79}$$

Since the freezing temperature is zero,

$$u(\underset{\sim}{x},\ell(\underset{\sim}{x})) = 0 \tag{4.80}$$

and, writing $S = \ell - t$, (3.8a) becomes

$$\nabla u(\underset{\sim}{x},\ell(\underset{\sim}{x})).\nabla\ell = L. \tag{4.81}$$

On the fixed boundary $\partial\Omega_1$, we prescribe the temperature

$$u = -\tilde{g}(\underset{\sim}{x},t) \le 0 \tag{4.82a}$$

and, if $\partial\Omega_2$ is non-empty,

$$u = 0 \quad \text{on} \quad \partial\Omega_2. \tag{4.82b}$$

Also the initial enthalpy and temperature distributions are

$$H_0(x) = \begin{cases} \alpha u_0(\underset{\sim}{x}), & \underset{\sim}{x} \in \Omega^-(0) \\ L \ , & \underset{\sim}{x} \in \Omega^0(0), \end{cases} \qquad u(\underset{\sim}{x},0) = \begin{cases} u_0(\underset{\sim}{x}), & \underset{\sim}{x} \in \Omega^-(0) \\ 0 \ , & \underset{\sim}{x} \in \Omega^0(0). \end{cases} \tag{4.83}$$

We have noticed earlier that the time derivative of the oxygen consumption problem is a one-phase Stefan problem but we may conversely proceed as in (2.58a,b) to define

$$U(x,t) = \begin{cases} \int_{\ell(\underset{\sim}{x})}^{t} u(\underset{\sim}{x},\tau)\,d\tau, & t > \ell(\underset{\sim}{x}), \quad \underset{\sim}{x} \in \Omega^0(0) & (4.84a) \\ 0 \quad , & t < \ell(\underset{\sim}{x}), \quad \underset{\sim}{x} \in \Omega^0(0) & (4.84b) \\ \int_0^t u(\underset{\sim}{x},\tau)\,d\tau \quad , & x \in \Omega^-(0) & (4.84c) \end{cases}$$

and show formally that

$$\alpha \frac{\partial U}{\partial t} = \nabla^2 U + H_0(x), \quad \underset{\sim}{x} \in \Omega_T^- \tag{4.85}$$

$$U = 0, \quad \underset{\sim}{x} \in \Omega_T^0 \tag{4.86}$$

$$U = 0, \quad \nabla U = \underset{\sim}{0}, \quad \underset{\sim}{x} \in \Gamma(t) \tag{4.87}$$

$$U = -\int_0^t \tilde{g}(\underset{\sim}{x},\tau)\,d\tau, \quad \underset{\sim}{x} \in \partial\Omega_1; \quad U = 0, \quad x \in \partial\Omega_2 \tag{4.88}$$

and

$$U(x,0) = 0, \quad x \in \Omega. \tag{4.89}$$

Hence, U formally satisfies the linear complementarity system

$$\alpha \frac{\partial U}{\partial t} - \nabla^2 U - H_0 \le 0, \quad U \le 0, \quad (\alpha \frac{\partial U}{\partial t} - \nabla^2 U - H_0)\,U = 0 \tag{4.90}$$

a.e. in Ω_T.

Comparing (4.90) with (4.76), we see that if $\alpha = 1$ and $\Omega^-(0)$ is empty so that $H_0(x) \equiv L$ in Ω, then $-U$ satisfies (4.76) with $c_0(\underset{\sim}{x}) = 0$, $L = R$ and $g(\underset{\sim}{x},t) = \int_0^t \tilde{g}(x,\tau)\,d\tau$. Let us now consider a class of parabolic variational inequalities of the type to which we will be able to show these complementarity systems are equivalent.

8. PARABOLIC VARIATIONAL INEQUALITIES

Let

$$K(t) = \{v \in H_0^1(\Omega): v \geq G(\underset{\sim}{x},t) \text{ a.e. on } \Omega\} \tag{4.91}$$

$$\mathcal{K} = \{v \in L^2(0,T,H_0^1(\Omega): v \geq G \text{ a.e. on } \Omega_T\} \tag{4.92}$$

where

$$G \in L^2(0,T;H^2(\Omega)), \quad \frac{\partial G}{\partial t} \in L^2(\Omega_T) \tag{4.93a}$$

and

$$G \leq 0 \text{ a.e. on } \partial\Omega \times (0,T). \tag{4.93b}$$

Also, suppose $f \in L^2(\Omega_T)$ and define, as usual,

$$(w,v) = \int_\Omega wv\, d\underset{\sim}{x}, \qquad a(w,v) = \int_\Omega \nabla w.\nabla v d\underset{\sim}{x}. \tag{4.94}$$

Then consider the following *parabolic variational inequalities*.

(P_t) Find $w \in L^2(0,T;H_0^1(\Omega))$ with $\partial w/\partial t \in L^2(0,T;L^2(\Omega))$

such that $w(\underset{\sim}{x},0) = w_0 \in K(0)$

and, for almost all $t \in (0,T)$, $w(t) \in K(t)$ and is such that

$$(\frac{\partial w}{\partial t}, v-w) + a(w,v-w) \geq (f,v-w)$$

for all $v \in L^2(0,T;H_0^1(\Omega))$ with $v(t) \in K(t)$ a.e. in $(0,T)$.

(P) Find $w \in \mathcal{K}$ with $\partial w/\partial t \in L^2(0,T;L^2(\Omega))$

such that $w(\underset{\sim}{x},0) = w_0(\underset{\sim}{x}) \in K(0)$

and $\int_0^T \{(\frac{\partial w}{\partial t}, v-w) + a(w,v-w) - (f,v-w)\}dt \geq 0$

for all $v \in \mathcal{K}$.

Theorem IV.8 [24, 215]

Problems (P) and (P_t) are equivalent and have a unique solution w such that $w \in L^2(0,T;H^2(\Omega))$.

Proof. (i) *Equivalence*

Clearly a solution of (P_t) solves (P). Conversely, consider a solution of

(P) and suppose, for any $\varepsilon > 0$ and $\tau \in (0,T)$,

$$v = \begin{cases} w \quad , & t \notin (\tau-\varepsilon,\tau+\varepsilon) \\ \tilde{v} \in K, & t \in (\tau-\varepsilon,\tau+\varepsilon) \end{cases} .$$

Then

$$\int_{\tau-\varepsilon}^{\tau+\varepsilon} \{(\frac{\partial w}{\partial t}, \tilde{v}-w) + a(w,\tilde{v}-w) - (f,\tilde{v}-w)\}dt \geq 0$$

and since this holds for any interval of integration in $(0,T)$ we obtain (P_t).

(ii) *Uniqueness*

Let w_i, $i = 1,2$ be two solutions of (P). Then, for $i \neq j$

$$\int_0^T \{(\frac{\partial w_i}{\partial t}, w_j-w_i) + a(w_i,w_j-w_i) - (f,w_j-w_i)\}dt \geq 0$$

so that, by addition,

$$\int_0^T \{\tfrac{1}{2} \frac{d}{dt}(w_1-w_2,w_1-w_2) + a(w_1-w_2,w_1-w_2)\}dt \leq 0$$

Thus, in the notation of p. 71,

$$\tfrac{1}{2}|(w_1-w_2)_T|_0^2 + \int_0^T |w_1-w_2|_1^2 \, dt \leq 0$$

which implies that $w_1 = w_2$.

(iii) *Existence*

Proceeding as for the obstacle problem, (4.24), we consider the nonlinear evolution equation

$$\frac{\partial w_\varepsilon}{\partial t} - \nabla^2 w_\varepsilon + \frac{1}{\varepsilon}(w_\varepsilon-G)^- = f, \quad \underset{\sim}{x} \in \Omega, \quad \varepsilon > 0 \tag{4.95a}$$

with

$$w_\varepsilon(\underset{\sim}{x},0) = w_0, \quad w_\varepsilon = 0 \text{ on } \partial\Omega. \tag{4.95b}$$

The nonlinearity penalises the violation of the constraint $w \geq G$. Since (4.95a) is an evolution equation for a monotone operator, (A), there exists a unique solution, [149]. Also it can be shown that $\|w_\varepsilon\|_{L^2(0,T;H^2(\Omega))}$, $\|\partial w_\varepsilon/\partial t\|_{L^2(\Omega_T)}$ and $\|(w_\varepsilon-G)^-/\varepsilon\|_{L^2(\Omega_T)}$ are uniformly bounded independently of ε, so that there is a subsequence w_ε, which converges weakly in $L^2(0,T;H^2(\Omega))$ and strongly in $L^2(\Omega_T)$ to a limit $\bar{w}_\varepsilon \geq G$ a.e. in Ω_T. Finally, multiplying (4.95a) by $v-w_\varepsilon$,

$$\int_0^T \{(\frac{\partial w_\varepsilon}{\partial t}, v-w_\varepsilon) + a(w_\varepsilon, v-w_\varepsilon) - (f, v-w_\varepsilon)\}dt \geq 0$$

for all $v \in K$, so that $\bar{w}$ solves (P) and $\bar{w} \in L^2(0,T;H^2(\Omega))$.

We can now prove the following theorem relating these parabolic variational inequalities to complementarity systems.

Theorem IV.9.

The solution to (P) satisfies the linear complementarity system

$$\frac{\partial w}{\partial t} - \nabla^2 w - f \geq 0, \quad w \geq G, \quad (\frac{\partial w}{\partial t} - \nabla^2 w - f)(w - G) = 0 \tag{4.96}$$

a.e. in Ω_T. The converse is also true.

Proof

For any non-negative $\phi \in C_0^\infty(\Omega_T)$, $v = w + \phi \in K$ and so, from (P)

$$\int_{\Omega_T} (\frac{\partial w}{\partial t} - \nabla^2 w - f)\phi \, d\underset{\sim}{x}dt = \int_0^T \{(\frac{\partial w}{\partial t}, v-w) + a(w,v-w) - (f,v-w)\}dt \geq 0$$

which implies that

$$\frac{\partial w}{\partial t} - \nabla^2 w - f \geq 0 \text{ a.e. in } \Omega_T.$$

Now let $\Omega_T^+ = \{(\underset{\sim}{x},t) \in \Omega_T : w(\underset{\sim}{x},t) > G(\underset{\sim}{x},t)\}$. Then for any $\phi \in C_0^\infty(\Omega_T^+)$, $v = w \pm \varepsilon\phi \in K$ for sufficiently small ε so that

$$0 = \int_0^T (\frac{\partial w}{\partial t}, \phi) + a(w,\phi) - (f,\phi)\,dt$$

$$= \int_{\Omega_T^+} (\frac{\partial w}{\partial t} - \nabla^2 w - f)\phi \, d\underset{\sim}{x}dt.$$

Hence,

$$\frac{\partial w}{\partial t} - \nabla^2 w - f = 0 \quad \text{a.e. in } \Omega_T^+$$

and (4.96) holds.

Conversely, by noting that if $w \in K$ satisfies (4.96), then, for $v \in K$,

$$(\frac{\partial w}{\partial t} - \nabla^2 w - f)(v - w) \geq 0 \quad \text{a.e. in } \Omega_T,$$

it is clear that w solves (P).

We can now trivially relate the oxygen consumption and one-phase Stefan problems to (4.96) and hence reformulate them as variational inequalities (P) or (P_t). We simply set $c = w-G$ in (4.76), where

$$\frac{\partial G}{\partial t} = \nabla^2 G + R + f \text{ in } \Omega_T \tag{4.97}$$

$$G = -g \text{ on } \partial\Omega_1 \tag{4.98a}$$

$$G = 0 \text{ on } \partial\Omega_2 \tag{4.98b}$$

and, when $\alpha = 1$, we set $U = G-w$ in (4.90), where

$$\frac{\partial G}{\partial t} = \nabla^2 G + H_0 + f \tag{4.99}$$

$$G = -g \text{ on } \partial\Omega_1 \tag{4.100a}$$

$$G = 0 \text{ on } \partial\Omega_2. \tag{4.100b}$$

The one-dimensional oxygen consumption problem was originally posed in [72] and has been studied in detail in [16]. We note that the model obstacle problem (2) of section 6 above is the steady state of an oxygen consumption problem in which the oxygen concentration is prescribed on the boundary of a half-plane [25].

The use of (4.84) to transform the one-phase Stefan problem to the variational inequality (P) was proposed in [89] and has been studied in more detail in [116]. Also, regularity results for (P) have been given in [24, 34, 115, 215].

We note that the variational inequality formulation for both these problems caters for the possibility that at some time $\Gamma(t)$ may intersect the fixed boundary $\partial\Omega_2$. However, this is a situation where the classical formulations are much harder to analyse rigorously than the variational ones and the relationship between the moving boundary problems (4.71a-4.74c, 4.79-4.83) and (P) appears to be an open question.

We conclude this section with a brief discussion of the Hele-Shaw class of problems introduced on pp. 22-23, which are a special case of (4.79-4.83). Considering (4.90) when $\alpha = 0$, so that $H_0 = 0$ in $\Omega^-(0)$ and $H_0 = L$ in $\Omega^0(0)$, we see from Theorem IV.4 and (4.56, 4.57) that U also solves the elliptic variational inequality in which $U \in K(t)$ and

$$a(U,v-U) \geq (H_0,v-U) \text{ for all } v \in K(t) \tag{4.101}$$

where

$$K(t) = \{v \in H^1(\Omega): v = -g(t) \text{ on } \partial\Omega_1,\ v = 0 \text{ on } \partial\Omega_2,\ v \leq 0 \text{ a.e. in } \Omega\}.$$

In this case the transformation (4.84) has the interesting property of relegating the role of the variable t to that of a parameter. The moving boundary at any particular time may be obtained simply by solving (4.101) at that time. In particular the Hele-Shaw problem described on p. 57, in which liquid is being injected at the constant rate $\partial p/\partial n = Q$ on $\partial\Omega_1$ and and $\partial p/\partial n = 0$ on $\partial\Omega_2$ becomes, after the transformation (4.84), the variational inequality (2.62) where

$$U,\ v \in K = \{H^1(\Omega): v \geq 0 \text{ a.e. in } \Omega\}.$$

There exists a unique solution for this Hele-Shaw problem for $t < T_0$, where T_0 is the time for the cell to fill i.e. for liquid to reach all $\partial\Omega_2$, [97].. The non-existence of a solution for $t > T_0$ can be regarded as a consequence of the non-coercivity of $a(v,v) = \int_\Omega |\nabla v|^2 d\underset{\sim}{x}$ in $H^1(\Omega)$. This means theorem IV.8 does not apply but more generally reflects that we have Neumann boundary data for a Poisson equation. It is generally true that for elliptic variational inequalities, the existence of a solution depends on the boundary and source data satisfying an integral *inequality*, [91, 150] whereas the inhomogeneous Neumann problem for a Poisson equation only has a solution if the data satisfies a certain integral *identity*.

Further numerical and mathematical studies of (2.62) may be found in [92, 96, 98, 110, 126].

On p. 24, we noted the relationship between these "Hele-Shaw" problems and limiting one-phase Stefan problems with large latent heat or small specific heat. It is possible to study the convergence of solutions of (4.90) to the solution of (4.101) in these limits and from the practical viewpoint, (4.101) generally provides a good approximation to the solution of the Stefan problem as long as the solidification does not continue for too long. (4.101) may clearly be solved more efficiently than (4.90).

9. NUMERICAL SOLUTION OF PARABOLIC VARIATIONAL INEQUALITIES

We will now consider finite difference and finite element approximations to the problems (P) and (P_t) of section 8, with $G \in C(\bar{\Omega}_T) \cap C[0,T;H^1(\Omega)]$.

(i) *Finite Difference Approximations*

For simplicity, we assume that Ω_T may be covered by a uniform rectangular net. Then the finite difference approximation to (4.96) is obtained by replacing spatial derivatives with central differences and the time derivative with a one-sided difference. The discrete problem is to find the vector $\underset{\sim}{w}^n$, which approximates w at the mesh points at the nth time step, such that

$$\underset{\sim}{Z}^{n+1} \geq \underset{\sim}{0}, \quad \underset{\sim}{w}^{n+1} \geq \underset{\sim}{G}^{n+1}, \quad (\underset{\sim}{Z}^{n+1})^T(\underset{\sim}{w}^{n+1} - \underset{\sim}{G}^{n+1}) = 0 \tag{4.102a}$$

where

$$\underset{\sim}{Z}^{n+1} = (\underset{\sim}{w}^{n+1} - \underset{\sim}{w}^n)/k + A\underset{\sim}{w}^{n+\theta} - \underset{\sim}{f}^{n+\theta} \tag{4.102b}$$

and where k, $\underset{\sim}{w}^{n+\theta}$, A and $\underset{\sim}{f}^{n+\theta}$ are as defined in (3.22). $\underset{\sim}{G}^{n+1}$ is the vector of values of G((n+1)k) at the mesh points. A simple example is, in one space dimension

$$\Omega = (0,1), \quad h = 1/m, \quad i \in [0,m] \tag{4.103a}$$

$$Z_i^{n+1} = \frac{1}{k}(w_i^{n+1} - w_i^n) + \frac{1}{h^2}(2w_i - w_{i-1} - w_{i+1})^{n+\theta} - f_i^{n+\theta} \tag{4.103b}$$

$$Z_i^{n+1} \geq 0, \quad w_i^{n+1} \geq G_i^{n+1}, \quad Z_i^{n+1}(w_i^{n+1} - G_i^{n+1}) = 0, \quad i \in (0,m) \tag{4.103c}$$

with

$$w_0^{n+1} = w_m^{n+1} = 0. \tag{4.103d}$$

(ii) *A Finite Element Approximation*

Let Ω be a polygonal domain in $\mathbb{R}^2$. Then, as on p. 70, we consider a finite element approximation of $H_0^1(\Omega)$ by the space V^h of continuous functions which are linear on each element of a triangulation $\mathcal{T}^h$ and which vanish on $\partial\Omega$. In a similar manner to our treatment of the obstacle problem on p. 108, we construct the following finite element discretisation of (P_t). Find $w^{n+1} \in K_h^{n+1}$ such that

$$((w^{n+1} - w^n)/k, v - w^{n+1})_h + a(w^{n+\theta}, v - w^{n+1}) \geq (f^{n+\theta}, v - w^{n+1})_h \tag{4.104}$$

for all $v \in K_h^{n+1} = \{v \in V^h : v_i \geq G(\underset{\sim}{x}_i, (n+1)k) \text{ for all interior nodes } i\}$.

(4.105)

The data $f^{n+\theta}$ and w^0 may be defined as functions in V^h such that

$$\left.\begin{array}{ll} \text{either} & f^n = \{f(nk)\}_I, \text{ if } f \text{ is continuous,} \\ \text{or} & (f^{n+\theta},v)_h = \frac{1}{k}\int_{nk}^{(n+1)k} (f(t),v)\,dt \text{ for all } v \in V^h \end{array}\right\} \tag{4.106a}$$

and

$$\left.\begin{array}{ll} \text{either} & w^0 = (w_0)_I, \text{ if } w_0 \in H^2(\Omega), \\ \text{or} & a(w^0,v) = a(w_0,v) \text{ for all } v \in V^h. \end{array}\right\} \tag{4.106b}$$

In any event

$$f_{h,k}(t) \to f \text{ in } L^2(\Omega_T), \quad w^0 \to w_0 \text{ in } H_0^1(\Omega) \tag{4.107}$$

where $f_{h,k}(t)$ is defined in $t \in (nk,(n+1)k)$ to be $f^{n+\theta}$.

The inner product $(.,.)_h$ may be either the L^2 inner product or the lumped mass inner product (3.26).

In exactly the same way, we could consider a more general convex region Ω with a smooth boundary in which an approximating polygon Ω^h may be inscribed. Then V^h is constructed as an approximation to $H_0^1(\Omega^h)$; the analysis is easier if we additionally require functions in K_h^{n+1} to vanish on the "skin" $\Omega\backslash\Omega^h$.

At each time step the solution is advanced by solving a discrete elliptic variational inequality which is equivalent to the linear complementarity problem

$$\underset{\sim}{Z}^{n+1} \geq \underset{\sim}{0}, \quad \underset{\sim}{w}^{n+1} - \underset{\sim}{G}^{n+1} \geq \underset{\sim}{0}, \quad (\underset{\sim}{Z}^{n+1})^T(\underset{\sim}{w}^{n+1} - \underset{\sim}{G}^{n+1}) = 0 \tag{4.108a}$$

where

$$\underset{\sim}{Z}^{n+1} = M(\underset{\sim}{w}^{n+1} - \underset{\sim}{w}^n)/k + \hat{A}\,\underset{\sim}{w}^{n+\theta} - \underset{\sim}{f}^{n+\theta}, \tag{4.108b}$$

$$M_{ij} = (b_i,b_j)_h, \quad \hat{A}_{ij} = a(b_i,b_j) \tag{4.108c}$$

and $\underset{\sim}{w}^n$, $\underset{\sim}{f}^n$ and $\underset{\sim}{G}^n$ are vectors of nodal values.

Since the matrices $I + \theta kA$ and $M + \theta k\hat{A}$ are symmetric and positive definite, (4.102) and (4.108) have unique solutions (p. 100) and these solutions may be obtained as on p. 111.

It is easily seen that when $\theta = 0$, the solution to (4.102) is

$$w_i^{n+1} = \max\{(\underset{\sim}{w}^n - kA\underset{\sim}{w}^n + k\underset{\sim}{f}^n)_i, G_i^{n+1}\} \tag{4.109}$$

which is simply the explicit scheme for advancing the equation $\partial w/\partial t - \nabla^2 w - f = 0$ followed by a projection onto the positive region of $\mathbb{R}^n$. This is also the case for (4.108) when **M** is the diagonal lumped mass matrix and in both cases the procedure is a version of the *truncation method*, [25]. In general the finite difference scheme for this method is to step from time level n to n+1 by solving

$$(\underset{\sim}{w}_*^{n+1} - \underset{\sim}{w}^n)/k + \theta A\underset{\sim}{w}_*^{n+1} + (1-\theta)A\underset{\sim}{w}^n = \underset{\sim}{f}^{n+\theta} \tag{4.110a}$$

followed by the truncation

$$\underset{\sim}{w}^{n+1} = \max(\underset{\sim}{w}_*^{n+1}, \underset{\sim}{G}^{n+1}). \tag{4.110b}$$

We observe that unless $\theta = 0$, the solution of (4.110) will not satisfy a linear complementarity system, although it will satisfy the constraint $\underset{\sim}{w}^{n+1} \geq \underset{\sim}{G}^{n+1}$.

(iii) *Stability*

As in Chapter III, we use the notation $v = \underset{\sim}{v}$, $\langle u,v\rangle = \underset{\sim}{u}^T M\underset{\sim}{v}$, $|v|^2 = \langle v,v\rangle$, $\|v\|^2 = \underset{\sim}{v}^T MA\underset{\sim}{v}$ and $\delta_k v^n = (v^{n+1} - v^n)/k$, with the convention that M = I for the finite difference approximation and $MA = \hat{A}$ for the finite element case. Also we find it convenient to assume

$$K(t) \supset K(t^*) \text{ for } t > t^* \text{ and } 0 \in K(0) \tag{4.111a}$$

or equivalently that

$$G(\underset{\sim}{x},t) \leq G(\underset{\sim}{x},t^*) \text{ for } t > t^* \text{ and } 0 \geq G(\underset{\sim}{x},0). \tag{4.111b}$$

Recalling from (4.15a) that the complementarity problems (4.102) and (4.108) are equivalent to

$$(\underset{\sim}{z}^{n+1})^T(\underset{\sim}{v} - \underset{\sim}{w}^{n+1}) \geq 0 \text{ for all } \underset{\sim}{v} \geq \underset{\sim}{G}^{n+1}, \tag{4.112}$$

and since, by (4.111), we may take $\underset{\sim}{v} = \underset{\sim}{w}^n$ in (4.112),

$$|\delta_k w^n|^2 + \langle Au^{n+\theta}, \delta_k w^n\rangle \le \langle f^{n+\theta}, \delta_k w^n\rangle \tag{4.113}$$

which is identical to (3.55). The argument used after (3.55) then gives the following stability theorem.

Theorem IV.10

Providing the stability condition

$$(1-2\theta)\ k[S(h)]^2 < 2 \tag{4.114}$$

holds when $\theta < \frac{1}{2}$, there is a constant c, independent of h and k such that

$$\sum_{j=0}^{n} k|\delta_k w^j|^2 + \|w^{n+1}\|^2 \le c\{\|w^0\|^2 + \sum_{j=0}^{n} k|f^{j+\theta}|^2\}, \tag{4.115a}$$

and so the finite element approximation satisfies

$$\sum_{j=0}^{n} \left|\frac{w^{n+1}-w^n}{k}\right|_0^2 + |w^{n+1}|_1^2 \le c\{|w^0|_1^2 + \sum_{j=0}^{n} k|f^{j+\theta}|_0^2\} \tag{4.115b}$$

Under certain conditions, the explicit method is stable in the sense of (4.115) when equality holds in (4.114). For example, consider the explicit method

$$\underset{\sim}{w}^{n+1} = \underset{\sim}{G}^{n+1} + \{(I-kA)\underset{\sim}{w}^n - k\underset{\sim}{f}^n - \underset{\sim}{G}^{n+1}\}^+ .$$

Lemma

If $1-k\,A_{ii} \ge 0;\quad A_{ij} \le 0,\ i \ne j;\quad \sum_j A_{ij} \ge 0$ (4.116a)

and

$$f(\underset{\sim}{x},t) = f(\underset{\sim}{x}) \le 0,\quad G(\underset{\sim}{x},t) = G(t) \le 0 \tag{4.116b}$$

with

$$G'(t) \le 0,\quad G''(t) \le 0 \text{ and } w(\underset{\sim}{x},0) = G(0) = 0 \tag{4.116c}$$

then $\{w^n\}$ is bounded by

$$\underset{\sim}{G}^{n+1} - \underset{\sim}{G}^{n} \le \underset{\sim}{w}^{n+1} - \underset{\sim}{w}^{n} \le \underset{\sim}{0}. \tag{4.117}$$

Proof

We proceed by induction. When $n = 0$, $\underset{\sim}{w}^1 - \underset{\sim}{w}^0 = \underset{\sim}{w}^1 \ge \underset{\sim}{G}^1 - \underset{\sim}{G}^0$. Also either $w_i^1 = G^1 \le 0 = w_i^0$ or

$$w_i^1 = \{(I-kA)\underset{\sim}{w}^0 + k\underset{\sim}{f}^0\}_i \le 0 = w_i^0.$$

Now suppose (4.117) is true for $n = 0,1,\ldots,m-1$. If $w_i^m = G_i^m$, then $w_i^{m+1} - w_i^m \ge G_i^{m+1} - G_i^m$ but otherwise

$$\begin{aligned} w_i^{m+1} - w_i^m &\ge \{(I-kA)(\underset{\sim}{w}^m - \underset{\sim}{w}^{m-1})\}_i \\ &\ge \{(I-kA)(\underset{\sim}{G}^m - \underset{\sim}{G}^{m-1})\}_i \\ &\ge G^m - G^{m-1} \text{ from (4.116a)} \\ &\ge G^{m+1} - G^m \text{ from (4.116b)}. \end{aligned}$$

If $w_i^{m+1} = G_i^{m+1}$ then $w_i^{m+1} \le G_i^m \le w_i^m$, but otherwise

$$w_i^{m+1} - w_i^m \le \{(I-kA)(\underset{\sim}{w}^m - \underset{\sim}{w}^{m-1})\}_i \le 0.$$

If (4.117) holds, then $\Sigma_j\, k|\delta_k w^j|^2$ is bounded by $\Sigma_j\, k|\delta_k G^j|^2$ and the proof of (3.58) shows that for (4.115a) to hold,

$$k \le \min(1/A_{ii},\ 2/[S(h)]^2) = h^2/4 \tag{4.118}$$

for the five-point formula for approximating $-\nabla^2$.

(iv) *Convergence*

To consider the convergence of the finite element discretisation, we define mesh functions on $(0,T)$ by, (cf. 3.59),

$$w_{h,k}(t) = w^{n+1}, \quad \hat{w}_{h,k}(t) = \{[(n+1)k-t]w^n + (t-nk)w^{n+1}\}/k$$

$$f_{h,k}(t) = f^{n+\theta}$$

for $t \in (nk,(n+1)k]$. The stability theorem IV.10 and (4.107) imply that the sequences $\{w_{h,k}\}$ and $\{\hat{w}_{h,k}\}$ are uniformly bounded in $L^2(0,T;H_0^1(\Omega))$ and $\{\hat{w}_{h,k}\}$ is uniformly bounded in $H^1(\Omega_T)$. Hence by compactness, there exist subsequences, which we do not distinguish notationally from the original sequences, such that

$$w_{h,k} \rightharpoonup w \text{ in } L^2(0,\tau;\ H_0^1(\Omega)) \tag{4.119a}$$

$$\hat{w}_{h,k} \rightharpoonup w \text{ in } H^1(\Omega_T) \tag{4.119b}$$

$$\hat{w}_{h,k} \to w \text{ in } L^2(\Omega_T). \tag{4.119c}$$

The strong convergence (4.119c) follows as in (3.60e). That the limits of the subsequences $w_{h,k}$ and $\hat{w}_{h,k}$ are the same follows from showing that $\lim\limits_{h,k\to 0} \int_{\Omega_T} (w_{h,k}-\hat{w}_{h,k})\phi d\underset{\sim}{x}dt = 0$ for all $\phi \in L^2(\Omega_T)$ and this is a consequence of the inequality

$$\left|\int_{\Omega_T} (w_{h,k}-\hat{w}_{h,k})\phi d\underset{\sim}{x}dt\right| \le k\left(k \sum_{n=0}^{N-1} \left|\frac{w^{n+1}-w^n}{k}\right|_0^2\right)^{\frac{1}{2}} \|\phi\|_{L^2(\Omega_T)}$$

(which itself follows from the Schwarz inequality) and the stability estimate (4.115).

To prove that w satisfies a parabolic variational inequality, we first note that for any $v \in C[0,T;H_0^1(\Omega)] \cap K$, there exists a $v_{h,k}(t) \to v$ in $L^2(0,T;H_0^1(\Omega))$ such that

$$v_{h,k}(t) = v^{n+1} \in K_h^{n+1} \text{ for } t \in (nk,(n+1)k]. \tag{4.120}$$

(For example, we could let v^{n+1} satisfy $a(v^{n+1}-\frac{1}{k}\int_{nk}^{(n+1)k} v(t)dt,\ v^*-v^{n+1}) \ge 0$ for all $v^* \in K_h^{n+1}$). When we substitute such a v into (4.104) and sum, we obtain,

$$k \sum_{n=0}^{N-1} \left(\frac{w^{n+1}-w^n}{k}, v^{n+1} - w^{n+1}\right)_h + a(w^{n+\theta},v^{n+1}-w^{n+1}) - (f^{n+\theta},v^{n+1}-w^{n+1})_h \ge 0$$

so that

$$\int_0^T \left(\frac{\partial \hat{w}_{h,k}}{\partial t}, v_{h,k}-w_{h,k}\right)_h + a(w_{h,k}, v_{h,k}-w_{h,k}) - (f_{h,k}, v_{h,k}-w_{h,k})_h dt$$

$$\geq (1-\theta)k^2 \sum_{n=0}^{N-1} a\left(\frac{w^{n+1}-w^n}{k}, v^{n+1}-w^{n+1}\right) = \chi_{h,k}, \quad (4.121)$$

say. Then, using the Cauchy-Schwarz inequality and (3.28),

$$|\chi_{h,k}| \leq k^2(1-\theta)\left(\sum_{n=0}^{N-1}\left|\frac{w^{n+1}-w^n}{k}\right|_1^2\right)^{\frac{1}{2}}\left(\sum_{n=0}^{N-1}|v^{n+1}-w^{n+1}|_1^2\right)^{\frac{1}{2}}$$

$$\leq (1-\theta)kS(h)\left(\sum_{n=0}^{N-1} k\left|\frac{w^{n+1}-w^n}{k}\right|^2\right)^{\frac{1}{2}}\left(\sum_{n=0}^{N-1} k|v^{n+1}-w^{n+1}|_1^2\right)^{\frac{1}{2}}$$

$$\leq \text{const. } k^{\frac{1}{2}}$$

provided $k(S(h))^2 \leq$ const., because of (4.115b) and the strong convergence of $v_{h,k}(t)$, (4.120). Hence, taking the limit in (4.121) and recalling (4.119), (4.120) and (4.106) and noting that the weak convergence of $w_{h,k}$ implies that

$$\liminf_{h,k\to 0} \int_0^T a(w_{h,k}, w_{h,k})dt \geq \liminf_{h,k\to 0} \int_0^T \{2a(w_{h,k}, w) - a(w,w)\}dt$$

$$= \int_0^T a(w,w)\,dt,$$

we finally obtain both the inequality

$$\int_0^T \left\{\left(\frac{\partial w}{\partial t}, v-w\right) + a(w, v-w) - (f, v-w)\right\}dt \geq 0 \quad (4.122)$$

for all $v \in C[0,T;H_0^1(\Omega)] \cap K$ and the initial condition $w(0) = w_0$.

In order to consider this inequality for all $v \in K$, we note that for such v, there exists a sequence

$$v_j \in C[0,T;H_0^1(\Omega)] \cap K \text{ such that } v_j \to v \text{ in } L^2(0,T;H_0^1(\Omega)],$$

[149, p. 269]. Thus (4.122) is satisfied for $v \in K$ and w is a solution of (P). The uniqueness of this solution implies that the limits of the sequences $w_{h,k}$, $\hat{w}_{h,k}$ are independent of the subsequences and that the sequences themselves converge. We thus have the following theorem.

Theorem IV.11

Let $f \in L^2(\Omega_T)$, $G \in L^2(0,T;H^2(\Omega)) \cap C[0,T;H^1(\Omega)] \cap C(\bar{\Omega}_T)$, $\partial G/\partial t \in L^2(\Omega_T)$, $G \le 0$ almost everywhere on $\partial\Omega \times (0,T)$ and $w_0 \in K(0)$. Then there exists a unique solution to (P) to which the finite element approximation converges in the sense that

$$\hat{w}_{h,k} \to w \text{ in } L^2(\Omega_T) \text{ and } \hat{w}_{h,k} \rightharpoonup w \text{ in } H^1(\Omega_T)$$

provided (4.114) holds.

We remark that if we try to relax the condition (4.111) and simply assume $0 \in K(t)$, then the result of putting $v = 0$ in (4.112) is

$$|w^n|_0^2 + \sum_{j=0}^{n-1} |w^{j+1}-w^j|_0^2 + \sum_{j=0}^{n} k|w^j|_1^2 \le \text{const.}\{|w^0|_0^2 + k\sum_{j=0}^{n-1} |f^{j+\theta}|_0^2\} \tag{4.123a}$$

provided

$$(1-\theta)k(S(h))^2 < 1, \quad \theta \ne 1. \tag{4.123b}$$

This is a more severe restriction on the step length than (4.114) and, following [123], it may be shown that $w_{h,k}(t) \to w$ in $L^2(0,T;H_0^1(\Omega))$ where w solves

$$\int_0^T \{(\frac{\partial v}{\partial t}, v-w) + a(w,v-w) - (f,v-w)\}dt + \tfrac{1}{2}|v(0)-w(0)|_0^2 \ge 0 \tag{4.123c}$$

for all $v \in K(t)$ and $\partial v/\partial t \in L^2(0,T;L^2(\Omega))$, provided $k(S(h))^2 \to 0$.
is obtained from (P) by integrating by parts in time.

These convergence results are rather weak and we conclude this section by reporting some error estimates which can be obtained if we demand more regularity of the solution at the outset. If $w \in L^\infty(0,T,H^2(\Omega))$ and $\partial w/\partial t \in L^2(0,T;H_0^1(\Omega))$, then it is shown in [26 , 134, 215] that the error in the finite element approximation satisfies

$$|w(nk) - w^n|_0 + (\sum_{n=1}^{N} k|w(nk)-w^n|_1^2)^{\frac{1}{2}} \leq \text{const. } (k^{\frac{1}{2}}+h). \qquad (4.124a)$$

The finite difference approximation of the one-dimensional oxygen consumption problem has been studied in [17] and, if the data is sufficiently smooth,

$$\max_{x,t} |\bar{w}_{h,k}(t) - w(t)| \leq \text{const. } (h^2+k)^{1-\varepsilon} \qquad (4.124b)$$

for any $\varepsilon > 0$, where $\bar{w}_{h,k}(t)$ is a spline function constructed from the values w_i^{n+1} at the mesh points.

(v) Numerical Examples

(a). *The Oxygen-consumption problem*

We consider the following two variational inequalities for almost all $t \in (0,T)$. Find $w(t) \in K(t)$ such that $w(x,0) = w_0(x)$ and

$$\int_0^1 \{\frac{\partial w}{\partial t}(v-w) + \frac{\partial w}{\partial x}(\frac{\partial v}{\partial x} - \frac{\partial w}{\partial x}) + (v-w)\}dx \geq 0 \qquad (4.125)$$

for all $v \in K(t)$ where

(a) $K(t) = \{H^1(0,1): v \geq 0 \text{ on } (0,1);\ v(0) = e^{t-1}-t,\ v(1) = 0\}$ (4.126a)

and $w_0(x) = e^{x-1}-x$; (4.126b)

(b) $K(t) = \{H^1(0,1): v \geq 0 \text{ on } (0,1);\ v(1) = 0\}$ (4.127a)

and $w_0(x) = \frac{1}{2}(1-x)^2$. (4.127b)

Thus w satisfies the linear complementarity system

$$\frac{\partial w}{\partial t} - \frac{\partial^2 w}{\partial x^2} + 1 \geq 0, \quad w \geq 0, \quad (\frac{\partial w}{\partial t} - \frac{\partial^2 w}{\partial x^2} + 1)\, w = 0 \qquad (4.128)$$

almost everywhere in $(0,1)\times(0,T)$ and also the moving boundary problem

$$\frac{\partial w}{\partial t} = \frac{\partial^2 w}{\partial x^2} - 1, \quad 0 < x < s(t), \quad 0 < t < T \qquad (4.129a)$$

with

$$w(s(t),t) = \frac{\partial w}{\partial x}(s(t),t) = 0. \qquad (4.129b)$$

Problem (a) is a model devised in [25] as a test for the truncation method and it has the closed form solution

$$w = \begin{cases} e^{t+x-1} - x - t, & 0 < x < 1-t \\ 0 \quad , & 1-t < x \end{cases} , \tag{4.130}$$

and problem (b) is the oxygen-consumption problem posed in [72].

Discretising (4.125) or (4.128) on a uniform partition of (0,1) with mesh length $h = 1/m$, using the finite element method (4.104) with a lumped mass matrix or the finite difference method (4.102) leads to the same algebraic problem

$$z_i^{n+1} = \left(\frac{w_i^{n+1} - w_i^n}{k} - \frac{(w_{i-1} - 2w_i + w_{i+1})^{n+\theta}}{h^2} + 1\right) \geq 0, \; w_i^{n+1} \geq 0, \tag{4.131a}$$

$$w_i^{n+1} \; z_i^{n+1} = 0 \tag{4.131b}$$

with

(a) $i = 1,2,\ldots,m-1$

$$w_0^n = e^{nk-1} - nk, \; w_m^n = 0, \tag{4.132}$$

(b) $i = 0,1,\ldots,m-1$

$$w_1^n = w_{-1}^n, \quad w_m^n = 0. \tag{4.133}$$

There is an ℓ^n such that the discrete solution satisfies $w_i^n > 0$ for $i < \ell^n$ and $w_i^n = 0$ for $i \geq \ell^n$. Thus the discrete moving boundary is at $\ell^n h$, and this result can be improved by the quadratic extrapolation formula (4.66b).

For problem (a), the results of calculations made with (4.131) and the truncation method with $\theta = 1$ are given in Table 8. Since the initial data is smooth we have taken $w_i^0 = w_0(ih)$. L2ERR is the percentage relative discrete L^2 error $= 100\{\Sigma|w_i^n - w(ih,nk)|^2/\Sigma|w(ih,nk)|^2\}^{\frac{1}{2}}$. The table suggests that this error is $O(h^2+k)$ which supports (4.124b) for this smooth data. Although the truncation method is less accurate than the inequality method, the latter is in general slower than the former because solving a complementarity problem is more time-consuming than solving linear algebraic equations. However, in this example, the method (4.55) was used for the complementarity calculation. The errors in the approximate values for s(t) are of $O(h)$ for $h = 1/16, 1/32, 1/64, 1/128$.

TABLE 8

L2ERR and s(t) for problem (a). $k = 128h^2/25$

Inequality Method							Truncation Method	
s(t)			L2ERR					
t	s=1-t	h=1/16	h=1/16	h=1/32	h=1/64	h=1/128	h=1/32	h=1/64
0.08	0.92	0.9206	1.9×10^{-1}	4.9×10^{-2}	1.2×10^{-2}	3.0×10^{-3}	0.23	0.11
0.16	0.84	0.8406	2.9×10^{-1}	7.6×10^{-2}	1.9×10^{-2}	4.6×10^{-3}	0.60	0.20
0.24	0.76	0.7629	3.6×10^{-1}	8.7×10^{-2}	2.2×10^{-2}	5.3×10^{-3}	0.97	0.28
0.32	0.68	0.6836	4.0×10^{-1}	9.3×10^{-2}	2.3×10^{-2}	5.5×10^{-3}	1.35	0.36
0.40	0.60	0.5972	3.9×10^{-1}	8.7×10^{-2}	2.2×10^{-2}	5.3×10^{-3}	1.83	0.47
0.48	0.52	0.5205	3.7×10^{-1}	8.7×10^{-2}	2.0×10^{-2}	5.2×10^{-3}	2.38	0.63
0.56	0.44	0.4425	3.6×10^{-1}	7.3×10^{-2}	1.7×10^{-2}	5.0×10^{-3}	3.35	0.87

For problem (b) the results of calculations using (4.131) starting at $t = 0.025$ with the small-time solution, [72, 128]

$$w \sim \tfrac{1}{2}(1-x)^2 - 2(t/\pi)^{\frac{1}{2}}\exp(-x^2/4t) + x\,\mathrm{erfc}(x/2t^{\frac{1}{2}})$$

are given in Table 9 where they are compared with the results of [128]. The small-time solution is used because of the flux discontinuity at $x = t = 0$. The values of s(t) are fairly accurate and were obtained quite cheaply.

(b). *One phase Stefan problem*

We consider the two-dimensional melting of a solid just below its melting temperature $u = 0-$ by the application of a temperature $u = 1$ at its boundary. The solid initially occupies $\Omega = (0,1)\times(0,1)$. As in (4.84) we define $U(\underset{\sim}{x},t) = \int_0^t u(\underset{\sim}{x},\tau)\,d\tau$ so that

TABLE 9

Free boundary position s(t) for problem (b).

t	h=0.05, k=0.005	h=0.01, k=0.001	Ref[128]
0.04	0.9990	0.9991	0.9992
0.06	0.9912	0.9916	0.9918
0.10	0.9333	0.9343	0.9350
0.12	0.8749	0.8792	0.8792
0.14	0.8010	0.7987	0.7989
0.16	0.6831	0.6833	0.6834
0.18	0.5171	0.5018	0.5011
0.185	0.4487	0.4342	0.4334

$$U \geq 0, \quad \frac{\partial U}{\partial t} - \nabla^2 U + L \geq 0, \quad U\left(\frac{\partial U}{\partial t} - \nabla^2 U + L\right) = 0 \qquad (4.134)$$

almost everywhere in Ω with

$$U = t \text{ on } \partial\Omega \text{ and } U(\underset{\sim}{x},0) = 0 \text{ in } \Omega. \qquad (4.135)$$

Because of symmetry, we can compute on the triangular region $R = \{0 \leq x \leq y \leq 1\}$ with an explicit finite difference scheme or, equivalently, with an explicit finite element scheme on a triangulation defined by equally spaced lines parallel to the sides of R and using a lumped mass matrix. We thus write, for $0 \leq i \leq j \leq m = 1/h$,

$$U_{i,j}^{n+1} = \{U_{i,j}^{n}(1 - \frac{4k}{h^2}) + U_{i-1,j}^{n} + U_{i+1,j}^{n} + U_{i,j-1}^{n} + U_{i,j+1}^{n} - kL\}^{+} \qquad (4.136)$$

$$U_{m,j}^{n} = nk, \quad 0 \leq j \leq m. \qquad (4.137)$$

Writing $w = U-t$ and $w_{i,j}^n = U_{i,j}^n - nk$, then $w_{i,j}^n$ is the approximation to w defined by (4.102) or (4.108) and so the stability condition is $k \leq h^2/4$ since $S(h) = 2\sqrt{2}$ and (4.116) are satisfied.

In Tables 10, 11 the calculated intersections of the moving boundary with $y = 0$ and $y = x$ are plotted for $k = h^2/4$ and compared with the results of [71]. Typical discrete moving boundaries are plotted in Fig. 29.

TABLE 10

Intersection of moving boundary with $y = 0$

t	h = 1/80	h = 1/40	h = 1/20	Ref[71]
0.05	.770	.770	.776	.775
0.10	.673	.674	.680	.675
0.15	.600	.600	.601	.600
0.20	.537	.539	.539	.535
0.25	.481	.480	.484	.477
0.30	.426	.426	.431	.422
0.35	.373	.373	.377	.368
0.40	.321	.321	.323	.315
0.45	.267	.267	.269	.261
0.50	.209	.210	.208	.204

TABLE 11

Intersection of moving boundary with $y = x$

t	h = 1/80	h = 1/40	h = 1/20	Ref[71]
0.05	.735	.735	.728	.732
0.10	.625	.624	.625	.623
0.15	.541	.540	.537	.541
0.20	.470	.469	.470	.471
0.25	.407	.408	.406	.409
0.35	.298	.298	.296	.301
0.40	.249	.248	.247	.252
0.45	.201	.200	.199	.204
0.50	.154	.153	.152	.156

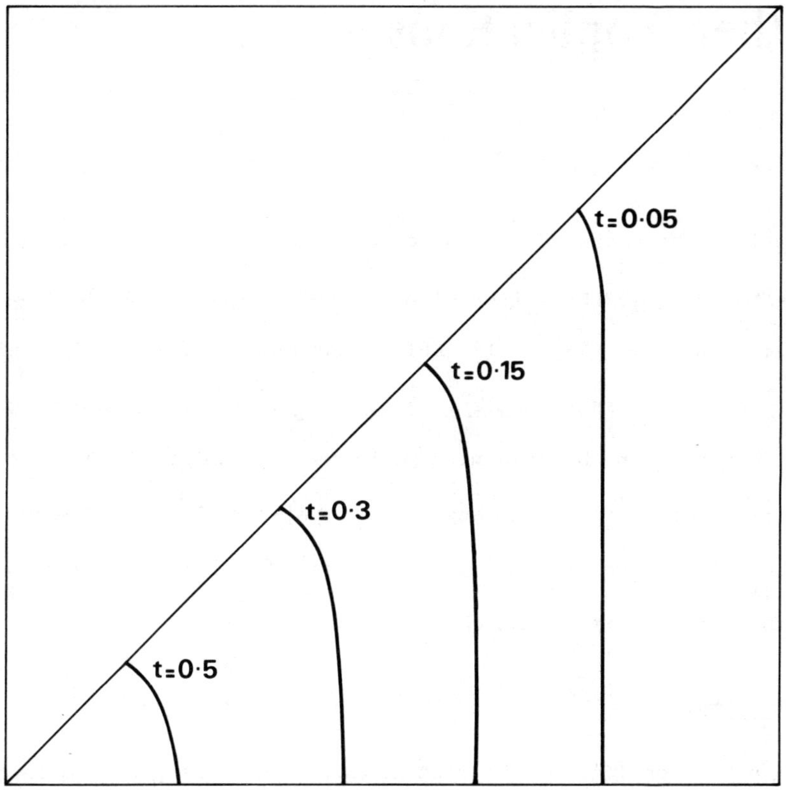

Fig. 29.

V Other Applications

In the previous three chapters we have described, at varying levels of rigour and completeness, the application of weak and variational methods to some of the problems listed in Chapter I. Before assessing the relative merits of these methods in the final chapter, we will first summarise the progress which can be made with the remaining problems of Chapter I. Our discussion will be non-rigorous and brief, and we will refer to published work for all proofs. We will however raise some open questions about generalised solutions in the second part of the chapter.

A. Contact Problems

Because the dimensionality of the free boundary in these problems is two less than that of the underlying field equations it is helpful to begin by considering a simpler free boundary problem (P) for a harmonic function u (Fig. 30) in a region Ω on part of whose boundary $\partial\Omega_3$, u is prescribed. On one part $\partial\Omega_1$, of the remainder of the boundary, $\partial u/\partial n = 0$ and, on the other part $\partial\Omega_2$, $u = g$, a prescribed function. The free boundary separates $\partial\Omega_1$ from $\partial\Omega_2$ and suppose u satisfies the minimum singularity condition that ∇u is continuous at this free boundary. In fact a version of this problem arises when we formally take the limit of the nearly rectangular dam problem (2.64, 2.65) when the upper surface of the dam is nearly the horizontal line $y = h_0$ and the reservoirs are full to within $O(\varepsilon\, h_0)$, $\varepsilon \ll 1$. When we write $p \sim -\rho g y + \varepsilon u$, $h \sim h_0 + \varepsilon g(x)$ and equate coefficients of ε, we retrieve a case of (P) where $\partial\Omega_1$ and $\partial\Omega_2$ are the projections on $y = h_0$ of the dry and wet parts of the upper surface of the dam. In this case we would know $u > g$ on $\partial\Omega_1$, $\partial u/\partial n \leq 0$

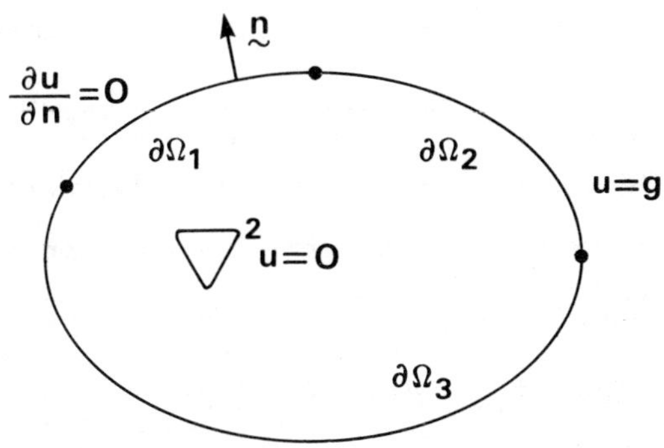

Fig. 30. A Model Free-boundary Problem (P)

on $\partial\Omega_2$ and that ∇u must be bounded at the free boundary.

Now suppose we try to minimise $\iint_\Omega |\nabla u|^2 \, dxdy$ over all sufficiently smooth functions u taking appropriate boundary values on $\partial\Omega_3$ and whose values on $\partial\Omega_1 \cup \partial\Omega_2$ are $\geq g$. Since the natural boundary condition for the minimisation of the unconstrained Dirichlet integral is $\partial u/\partial n = 0$, it is plausible that this is the condition which will be satisfied at those portions of $\partial\Omega_1 \cup \partial\Omega_2$ at which the condition $u = g$ fails to hold. We can also interpret this natural boundary condition physically if we regard (P) as the limit of an obstacle problem (1.10) for a membrane fixed along $\partial\Omega_3$ and unstressed along $\partial\Omega_1 \cup \partial\Omega_2$ when the obstacle $u = G$ both thins and approaches the boundary $\partial\Omega_1 \cup \partial\Omega_2$ (Fig. 31). Indeed this "thin obstacle problem" has been

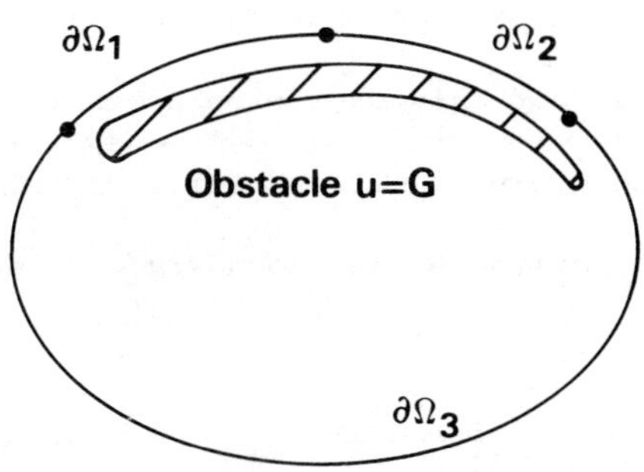

Fig. 31. Thin Obstacle Problem

considered theoretically in its own right [113], but a direct treatment of

(P) is also possible as a unilateral variational principle [107, 139].

The analysis of the zero-friction two-dimensional Signorini problem (p. 9) of contact between an elastic body and a rigid obstacle or support proceeds similarly. We recall that the stress $\underset{\sim}{T}$ is zero on those portions of the boundary of the elastic body not in contact with the support, and on these portions the displacements are such that $(u,v).\underset{\sim}{n} < d$ where the outward normal is $\underset{\sim}{n}$; there is a prescribed normal displacement $(u,v).\underset{\sim}{n} = d$, zero tangential stress $\underset{\sim}{T}.\underset{\sim}{t}$ and negative normal stress $\underset{\sim}{T}.\underset{\sim}{n}$ on those portions which are in contact. Also body forces are neglected and we assume prescribed displacements on the boundary of the elastic body away from the support for simplicity. Noting from (1.6, 1.7) that the stress is related to the energy density

$$E = G\{(\frac{\partial u}{\partial x})^2 + (\frac{\partial v}{\partial y})^2 + (\frac{\partial u}{\partial y} + \frac{\partial v}{\partial x})^2 + \frac{\lambda}{2}(\frac{\partial u}{\partial x} + \frac{\partial v}{\partial y})^2\} \tag{5.1}$$

by

$$\begin{pmatrix} \sigma_x & \tau \\ \tau & \sigma_y \end{pmatrix} = \begin{pmatrix} \partial E/\partial(\frac{\partial u}{\partial x}) & \partial E/\partial(\frac{\partial u}{\partial y}) \\ \partial E/\partial(\frac{\partial v}{\partial x}) & \partial E/\partial(\frac{\partial v}{\partial y}) \end{pmatrix} , \tag{5.2}$$

we simply minimise

$$\iint\limits_{\text{elastic body}} E\, dxdy \tag{5.3}$$

not over all (u,v) satisfying prescribed boundary data, as we would in a classical elasticity problem, but only over displacements which are such that $(u,v).\underset{\sim}{n} \le d$. The natural boundary conditions for this problem are $\underset{\sim}{T} = \underset{\sim}{0}$ when neither the tangential nor the normal boundary displacement is prescribed and in particular $(u,v).\underset{\sim}{n} < d$; when contact occurs and $(u,v).\underset{\sim}{n} = d$, $\underset{\sim}{T}.\underset{\sim}{n} < 0$, the natural boundary condition is $\underset{\sim}{T}.\underset{\sim}{t} = 0$. The proof that (5.3) is the correct unilateral variational problem is given in [107].

The extension of these ideas to frictional contact problems is still unclear, although certain results are known when the frictional constraint is replaced by $|\underset{\sim}{T}.\underset{\sim}{t}| \leq \mu T_{on}$, where T_{on} is a *prescribed* normal stress [187, 240].

B. Generalised Stefan Problems

As mentioned earlier, the theory of weak solutions for Stefan problems applies only in certain idealised situations in which several of the thermal parameters are restricted. Nonetheless there have been numerous attempts to extend the theory to more general or more physically realistic models. Volumetric or body heating is an interesting effect which is important in the theory of resistance welding [98, 202] and thermal switches [76], but we will postpone discussion of this until the end of the chapter. Another generalisation is that of alloy solidification, in which coupled heat and mass transfer occurs [27, 77, 86, 109, 144]. We will not even describe the simplest model here but simply point out that there is no obvious extension of the uniqueness result (2.34) to systems of parabolic equations.

One physical generalisation which can be approached by the methods of the Chapter III is that of ablation in one space dimension [31]. Here melting occurs through an energy source $Q(t) > 0$, arising perhaps from a high speed flow or a laser, which is strong enough to remove the liquid almost instantaneously either mechanically or by vapourisation. This situation may be modelled as a one-phase Stefan problem in $0 < x < s(t)$ with the energy balance

$$k \frac{\partial u}{\partial x} = \rho L \frac{ds}{dt} + Q \quad \text{at} \quad x = s-0. \tag{5.4}$$

We may embed this problem in a fixed domain by replacing the energy supply Q by a fictitions liquid phase in which

$$k \frac{\partial^2 u}{\partial x^2} = 0, \quad s(t) < x < d \tag{5.5}$$

where d is sufficiently large and

$$k \frac{\partial u}{\partial x} = Q(t) \quad \text{at} \quad x = d. \tag{5.6}$$

The boundary condition (1.15) for this two-phase Stefan problem is

$$[k \frac{\partial u}{\partial x}]_{\text{solid}}^{\text{liquid}} = -\rho L \frac{ds}{dt} = -k \frac{\partial u}{\partial x}\bigg|_{\text{solid}} + Q \tag{5.7}$$

from (5.5). Thus, if we define

$$H(u) = \begin{cases} \rho c u & u < 0 \\ \rho L & u > 0, \end{cases} \tag{5.8}$$

we can define a weak solution of the ablation problem which may be studied within the framework of Chapter III. However, the extension of this idea to general ablation problems in more space dimensions is not clear.

Although this is the only Stefan problem we shall consider in which fluid flow effects are taken into account, we will consider other problems in heat and mass transfer in which there is a prescribed convective term in the final chapter.

C. Irrotational Water Waves

Although this problem has a variational formulation, it is one of the problems mentioned in Chapter I with which least progress has been made. In the simplest case of steady flow above an impermeable base, we consider the stationary values of

$$\int_{x_1}^{x_2} \int \{\tfrac{1}{2}|\nabla\phi|^2 + gy\} dy dx \tag{5.9}$$

where x_i are arbitrary and the y integration is over the depth of the fluid, [152]. By the usual arguments, the Euler equation in the interior of the flow is Laplaces equation for ϕ and from (2.43) the contribution from the free boundary to the first variation is

$$\oint_{f=0} \{\frac{\partial\phi}{\partial n}\,\delta\phi - \frac{\delta f}{|\nabla f|}\,[\tfrac{1}{2}|\nabla\phi|^2 + gy]\}\,ds \qquad (5.10)$$

where $\delta\phi$ and δf are the variations in the potential and the free surface $f(x,y) = 0$. Hence, if we allow these variations to vary independently, we retrieve (1.23) and (1.25).

It is interesting to contrast (5.9) with the usual Lagrangian which is the difference between the kinetic and potential energies. We may use such a Lagrangian, but if we persist with ϕ as a dependent variable, then $\delta\phi$ and δf must not be allowed to vary independently. However, if we work in terms of the stream function ψ (p. 19), which we may take to be zero on $f = 0$, and replace the integrand of (5.9) by $\tfrac{1}{2}|\nabla\psi|^2 - gy$, we still obtain $\nabla^2\psi = 0$ in the interior of the flow, but

$$\oint_{f=0} \{\frac{\delta f}{|\nabla f|}\,(\frac{\partial\psi}{\partial n})^2 - \frac{\delta f}{|\nabla f|}\,[\tfrac{1}{2}|\nabla\psi|^2 - gy]\}\,ds \qquad (5.11)$$

replaces (5.10) since $\psi + \delta\psi = 0$ on $f + \delta f = 0$, which again leads to (1.23), (1.25).

While a numerical implementation of (5.9) is possible [2], no obvious fixed domain method suggests itself. The situation may be compared with the formulation of the transformed rectangular dam problem (2.67, 2.68) as finding the minimum of

$$\int_0^L \int_0^h \{\tfrac{1}{2}|\nabla U|^2 + \rho g U\}\,dx\,dy \qquad (5.12)$$

over all functions U which satisfy appropriate conditions at the fixed boundaries $x = 0, L$, $y = 0$. When δU and δh are allowed to vary with $U + \delta U = 0$ on $y = h + \delta h$, we find that $\partial U/\partial n$ vanishes on $y = h$. However unlike (5.9), (5.12) also has a fixed domain formulation as described in Chapter IV.

Time-dependent water wave problems may also be formulated as variational problems by adding $\partial\phi/\partial t$ to the integrand of (5.9) and integrating between two arbitrary times at which $\delta\phi$ and δf vanish. This formulation is useful in seeking invariants of the motion [23].

D. Lubrication Theory

The formulation (1.27) for the simplest one-dimensional model of lubrication theory, with the so-called Reynolds or Swift-Steiber boundary condition

$$p = 0, \ \frac{dp}{dx} = 0 \tag{5.13}$$

at the cavity boundary, is already suitable for treatment as a variational inequality. The linear complementarity problem is

$$p\left(\frac{d}{dx}\left(\frac{h^3}{6}\frac{dp}{dx}\right) - U\frac{dh}{dx}\right) = 0 \tag{5.14a}$$

with

$$p \geq 0, \quad \frac{d}{dx}\left(\frac{h^3}{6}\frac{dp}{dx}\right) - U\frac{dh}{dx} \leq 0. \tag{5.14b}$$

Hence, for a bearing in $-L \leq x \leq L$ with a prescribed pressure $p = p_0 > 0$ at $x = -L$, the variational inequality is

$$a(p, v-p) \geq \left(-U\frac{dh}{dx}, v-p\right) \tag{5.15a}$$

where

$$a(u,v) = \int_{-L}^{L} \frac{h^3}{6}\frac{du}{dx}\frac{dv}{dx}\,dx \tag{5.15b}$$

$$(u,v) = \int_{-L}^{L} uv\,dx \tag{5.15c}$$

and v is an arbitrary positive function which is equal to p_0 at the fixed boundary $x = -L$. The fact that (5.13) is satisfied by the solution of (5.15) under certain circumstances follows from theory of Chapter IV for the obstacle problem (p. 103). The possibility of formulating the free boundary problem (1.27, 5.13) as a variational inequality was observed in [47, 59, 206]. An earlier method, [55], for solving the free boundary problem as a form of projected relaxation was later justified in [79].

This is another good example for illustrating the importance of the second inequality in (5.14b) in ensuring that (5.14) has a unique solution.(cf. p. 93-94). The bearing in Fig. 32, as formulated as a free boundary problem can have lubricated regions OA or BC or both. However, the variational inequality or linear complementarity formulations will have the unique solution in which both regions are lubricated.

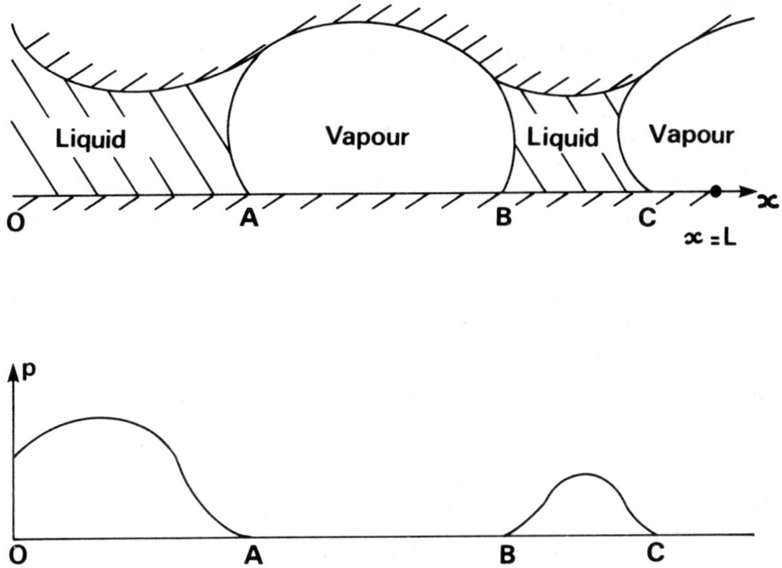

Fig. 32. Partially Lubricated Bearing

We should point out that several conditions different from (5.13) have been proposed to model cavitation in different types of bearing with different amounts of "leakage" of lubricant into the cavity [68, 87, 159, 189, 214]. Also instability of the cavity boundary may invalidate one-dimensional models [189, 214].

E. Immiscible Fluid Flow in Porous Media and Hele-Shaw Cells

As mentioned earlier, the porous medium problem (1.34)-(1.38) has provided one of the principle motivations and successes of the variational methods described in Chapter IV. It is thus surprising that the theory of the related problem of flow of two immiscible liquids in a porous medium remains much less clear, even in the absence of gravity. In this so-called "Muskat" problem [169], the liquid velocities $\underset{\sim}{q}_i$ satisfy, as in (1.34),

$$\underset{\sim}{q}_i = -K_i \nabla p_i \tag{5.16}$$

where K_i is inversely proportional to the liquid viscosity μ_i. As in (1.35) the incompressibility of the liquids implies that

$$\nabla^2 p_i = 0 \tag{5.17}$$

but continuity of momentum and mass at the interface between the liquids now requires that

$$p_1 = p_2, \quad K_1 \frac{\partial p_1}{\partial n} = K_2 \frac{\partial p_2}{\partial n} = -v_n. \tag{5.18}$$

Here, as earlier, v_n is the normal velocity of the interface with a suitable orientation. If this interface is written as $f(x,y,t) = 0$ in a two-dimensional flow, (5.18) becomes

$$\frac{\partial f}{\partial t} - K_i \nabla p_i . \nabla f = 0. \tag{5.19}$$

This model also describes the motion of immiscible liquids in a Hele-Shaw cell, but as far as generalised solutions are concerned, the problem remains intractable by the weak and variational methods described in this book. However, we may note that if the liquids have disparate viscosities so that $K_1 = 1$, $K_2 = \delta^{-1} \gg 1$ say, we may seek formal asymptotic extensions of the form $p_1 \sim p_{10} + \ldots$, $p_2 \sim \delta p_{20} + \ldots$

where

$$\nabla^2 p_{i0} = 0$$

and, from (5.18),

$$p_{10} = 0, \; -\frac{\partial p_{10}}{\partial n} = -\frac{\partial p_{20}}{\partial n} = v_{n0}. \tag{5.20}$$

Hence the problem for p_{10} uncouples from that for p_{20}, the latter being the solution of a Neumann harmonic problem once p_{10} is known. Moreover, p_{10} satisfies the Hele-Shaw model (1.28)-(1.31) and can, in suitable circumstances, be found by the methods of Chapter IV.

F. Stability and Mushy Regions

We have remarked in the introduction that while weak and variational methods provide powerful theoretical tools for the analysis and numerical solution of free and moving boundary problems, they can obscure some of the information which can be obtained by more classical methods, in particular by perturbation methods. Many useful results have been obtained by heuristic asymptotic methods in which either the space or time variables or one or more of the physical parameters are allowed to become large or small. We can illustrate the breadth of applicability of these methods by mentioning the local analyses of singularities, such as the top of the seepage face in the dam problem (2.64, 2.65), [223], small and large latent heat expansions of Stefan problems [125, 227], large time solutions of Stefan and percolation problems, [137, 209] and theories of small amplitude, or shallow, or weakly dispersive water waves, [228]. However, one particular asymptotic analysis may have implications for some of the weak and variational formulations described in this book and that is stability theory.

A very simple example of a local linear stability analysis of a moving boundary problem concerns Hele-Shaw flow with a smooth free boundary. Near this boundary we regard the liquid as occupying the entire half-space $x < Vt$, where V is a given constant and is the local normal velocity of the free boundary. The functions

$$p = V(Vt - x), \quad f = x - Vt \tag{5.21a,b}$$

satisfy (1.28)-(1.31) and we may seek a linearly perturbed solution in which the free boundary has an oscillation of small amplitude ε. We consider temporal stability by prescribing a positive wavenumber n, which must not be so large that slope variations at the free boundary are of $O(1)$, and then seeking an exponential growth or decay in time. Writing

$$f \sim x - Vt - \varepsilon \sin ny \, e^{\sigma t} + \ldots \tag{5.22}$$

$$p \sim V(Vt - x) + \varepsilon A \sin ny \, e^{nx+(\sigma-nV)t} + \ldots, \tag{5.23}$$

where A is to be determined, and equating coefficients of ε in (1.28)-(1.31) gives an eigenvalue problem for σ with solution

$$\sigma = -nV . \tag{5.24}$$

This equation for the growth or decay rate in terms of the given wavenumber is called a dispersion relation and it suggests that if $V > 0$ and the free boundary is advancing, all wavenumbers will be stable but if $V < 0$ and the boundary is receding, all wavenumbers will be unstable, the shortest wavelengths have the greatest growth rates. This latter result confirms the ill-posedness of the Hele-Shaw model for contracting regions mentioned on p. 24.

This stability result has a simple physical interpretation. If, in Fig. 33, the pressure is negative in the liquid so that the free boundary is receding, then the pressure at A will be less than it would be if the free boundary was straight. This will cause inflow so that the small protuberance near A will grow into a lengthening finger, but outflow and decay of the protuberance will occur when the free boundary is advancing.

This is one of the simplest stability analyses of free and moving boundary

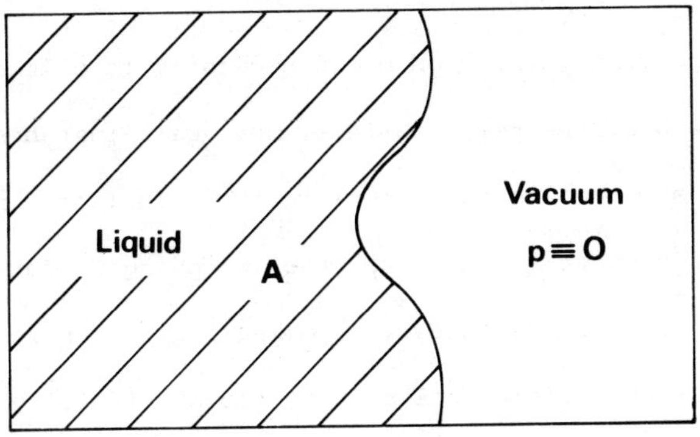

Fig. 33. Hele-Shaw Cell

problems and the effects of many other physical mechanisms on (5.24) have been studied, some of which are listed in the bibliography of [179]. For example, if a surface tension T in the (x,y) plane is taken into account, (5.24) becomes

$$\sigma = -nV - Tn^3 \tag{5.25}$$

so that when $V < 0$, there is a critical or most dangerous wavenumber $n = \sqrt{-V/T}$. The importance of such results in the theory of crystal growth is discussed in [144]. Again, if a gravity field $g(\cos\alpha, \sin\alpha)$ is important, (5.21a) becomes $p = (V - \rho g\cos\alpha)(Vt - x)$ so that

$$\sigma = -n(V - \rho g\cos\alpha). \tag{5.26}$$

Finally, if the Laplacian in (1.28) is replaced by the two-dimensional diffusion operator with thermal diffusivity κ, so that the problem becomes a one-phase Stefan problem with p being the liquid temperature, we find that if $V > 0$,

$$\sigma = -nV \quad \text{for} \quad n > V/\kappa \tag{5.27}$$

but when $V < 0$,

$$\sigma = -nV, \quad 0 < n < -V/\kappa \tag{5.28a}$$

and

$$\sigma = nV, \quad 0 < n \tag{5.28b}$$

are both possible. The restrictions on n are necessary to eliminate exponential spatial growth of the perturbations, but again instability can only occur when $V < 0$ and the liquid is supercooled. In fact a more complicated analysis can be given to show that the two-phase Stefan problem is only linearly unstable in the presence of superheating or supercooling, [239].

With considerably greater analytical effort, global stability analyses can be performed by perturbing about explicit similarity solutions with planar or spherical symmetry, [52, 211], and even more complicated local weakly nonlinear calculations have been performed for Stefan and alloy solidification problems, [144, 250].

Although the dispersion relations (5.24)-(5.28) have been derived by seeking elementary linearised solutions, they arise in the solution of initial value problems for data which differs by a small amount from data which corresponds to a trivial explicit solution. The dispersion relations then correspond to poles of the Fourier transforms of the solutions of these initial value problems, and it is in this context that they can be best understood. In any event, they suggest that one-phase problems of Hele-shaw or Stefan type are unstable when the moving boundary is receding and the field equation is to be solved in a contracting region and it is interesting that the few known non-trivial solutions to these linearly unstable problems possess other undesirable features. Finite time "blow-up" can occur in one-dimensional supercooled Stefan problems, [219] and in Hele-Shaw problems with suction, [199] and one parameter families of travelling wave solutions (p. 25) can also occur in the latter example [213]. Also the Neumann solution mentioned on p. 14 ceases to exist if $-L$ is too large.

When linear theory predicts instability, the traditional approach is either to try to take nonlinear effects into account or to revise the model by including some new stabilising mechanism, as in (5.25). Either or both these approaches would be natural if it was known from observation that planar free or moving boundaries were capable of developing non-planar "fingers" whose detailed shape was of interest. However, if we have a fixed domain numerical method for solving the problem, another possibility presents itself. It may happen that even though the classical statement of the problem leads to a local linear instability of the interface, the generalised solution to which the numerical discretisation tends may not exhibit such an unstable free or moving boundary. Before considering the physical interpretation of this possibility, let us consider one example where it has been observed to occur.

The Stefan problem

$$\frac{\partial u}{\partial t} = \frac{\partial^2 u}{\partial x^2} + Q, \qquad 0 < x < 1, \qquad x \neq s(t), \quad Q > 0, \tag{5.29}$$

with

$$u(s,t) = 0, \qquad \left[\frac{\partial u}{\partial x}\right]_{s+0}^{s-0} = -\rho L \frac{ds}{dt} \tag{5.30}$$

and

$$u(1,t) = -1, \qquad \frac{\partial u}{\partial x}(0,t) = 0, \qquad u(x,0) = -1 \tag{5.31}$$

is a possible model for the resistance welding of two metal plates, [9] (Fig. 34). The temperature is $u(x,t)$, Q is a simple representation of the resistance Joule heating and the electrodes are long enough compared to the plate thickness for a one-dimensional model to be used. No explicit solution for u is available but the temperature distribution at successive times is shown qualitatively in Fig. 35 (i), (ii), (iii). No melting takes place for $t < t_m$, where t_m is the time taken for $u(0,t)$ to reach the melting temperature zero. For $t > t_m$, the model necessitates the appearance of a phase boundary $x = s(t)$ with $s(t_m) = 0$. Now let us define $s(t_m+\delta t) = \delta s$,

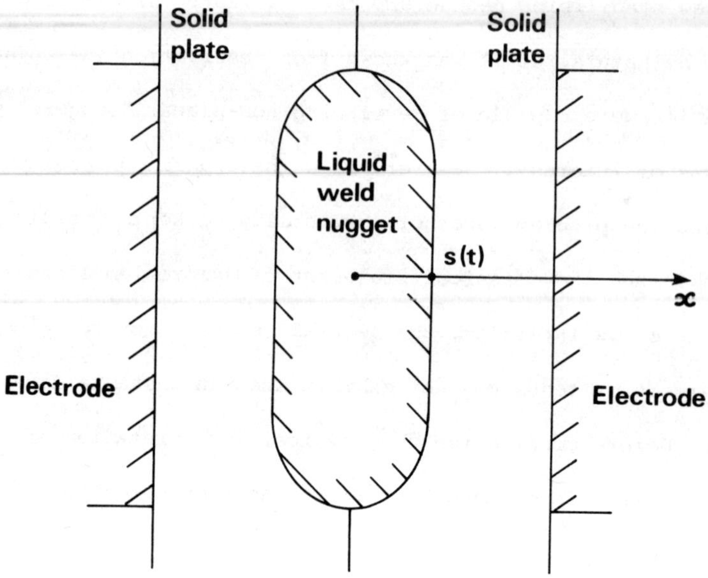

Fig. 34. Resistance Welding

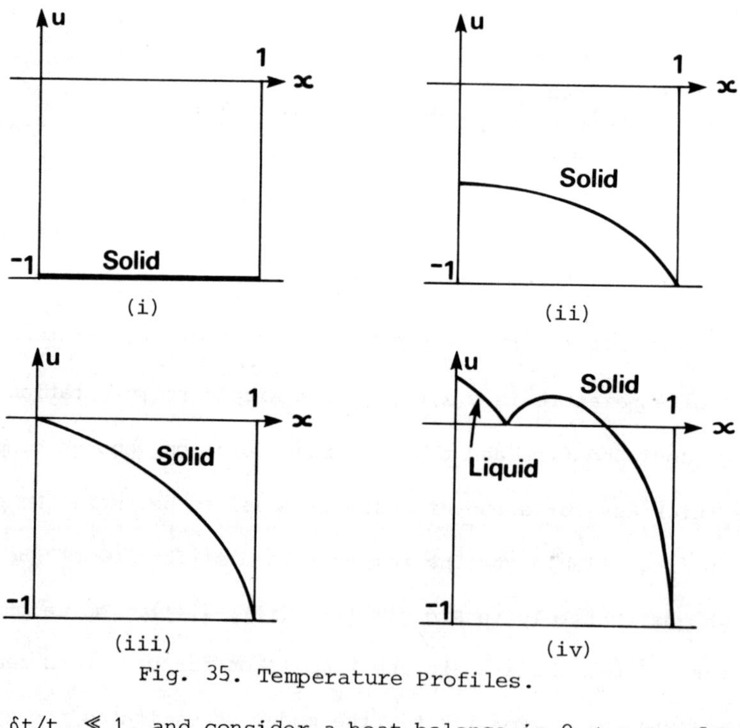

Fig. 35. Temperature Profiles.

where $0 < \delta t/t_m \ll 1$, and consider a heat balance in $0 < x < s$ for $t_m < t < t_m + \delta t$. The initial and final temperature profiles are as in Fig.

36. so the net heat gain exceeds $\rho L\ \delta s$. This heat originates partly from the volumetric heating, which accounts for $Q\delta s\delta t$ and partly from heat flow from the region $x > \delta s$. Thus, for sufficiently small δt, there must be a

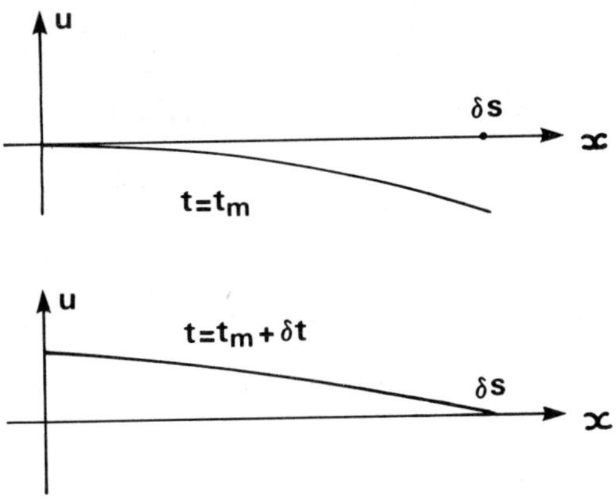

Fig. 36.

positive heat flow from the region $x > \delta s$ and this can only happen if the region $x > \delta s$ is superheated as in Fig. 35 (iv). Hence the phase boundary $x = s(t)$ will be unstable from our earlier discussion.

Now suppose the solution of this problem is computed by the enthalpy method of Chapter III [9, 202]. We know this gives the weak solution of (5.29)-(5.31), which may not coincide with the "superheated classical solution" described above. In fact a temperature distribution is found which is identical with Fig. 35 (i)-(iii) for $t \le t_m$ but which for $t > t_m$ predicts the "mushy region" shown in Fig. 37. The temperature is zero not on a curve in the (x,t) plane but over a region bounded by two curves $x = s_i(t)$ where $s_1(t_m) = s_2(t'_m) = 0$. In this region the enthalpy H varies with x and t, but u does not. Until the time t_I at which the curves $x = s_i(t)$ intersect, the classical energy balance (1.15) is satisfied at neither curve; the flux $-\partial u/\partial x$ is continuous, i.e. zero, at the solid-mushy boundary $x = s_1(t)$ and has a jump or finite discontinuity at the liquid-mushy boundary $x = s_2(t)$ of an amount needed to ensure that the total enthalpy change per unit length at any position x between times t_s and t_L (Fig. 37) is ρL. Moreover both the moving boundaries $x = s_i(t)$ are linearly stable on the basis of an analysis similar to that on p. 163. We also note that since the weak solution is also the limit of the solutions of a sequence of regularised problems with continuous enthalpy (3.13), we can say that, in the solutions to these regularised problems, the curves on which $H = 0, \rho L$ will be nearly at $x = s_i(t)$ even when the regularisation parameter ε is very small.

The physical existence of such a mushy region in which the liquid and solid phases coexist requires much more justification than is at present available. All we will do here is conjecture that such a region consists of many "blobs" of pure liquid and pure solid, the complicated blob geometry being the result of the instability caused by a slight superheating of the solid. Thus we may visualise an initially planar interface evolving as in Fig. 38 with the averaged or "homogenised" temperature in the region $s_2 < x < s_1$ being zero. By this we mean the temperature averaged over a region containing many blobs yet still small compared with the size of the mushy region. This concept occurs in alloy solidification where an impurity

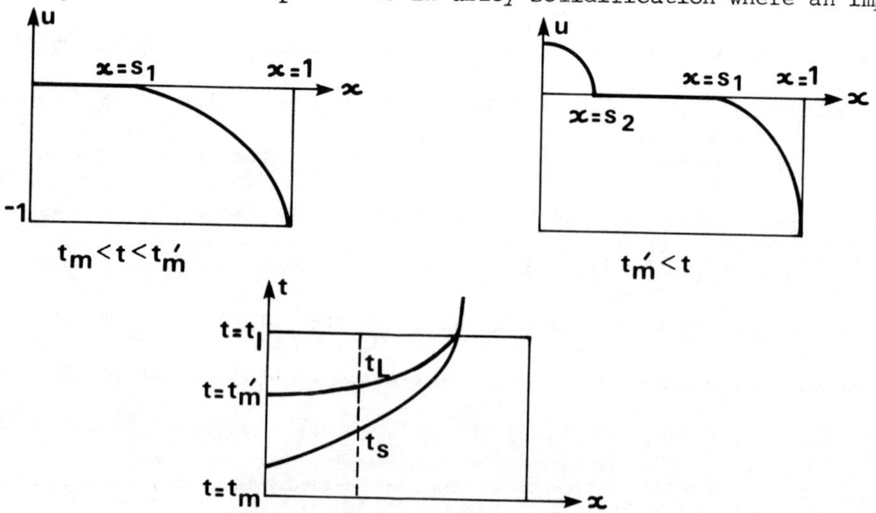

Fig. 37. Mushy Region

is diffusing through the liquid region and the mushy region contains dendrites of various types, [231].

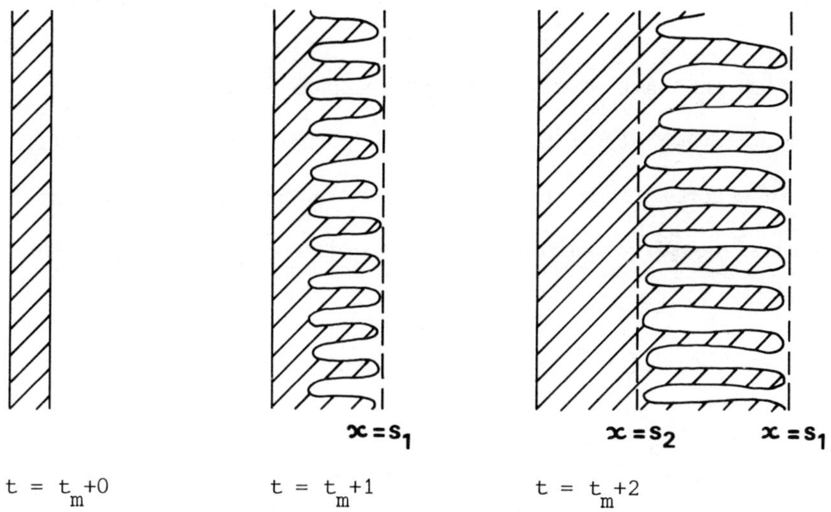

Fig. 38. Possible Development of Mushy Region
Shaded Region is Liquid

VI Generalisations and Conclusions

1 CONSERVATION LAWS AND FREE AND MOVING BOUNDARY PROBLEMS

The weak formulation of the Stefan problem considered in Chapter III was motivated in Chapter II by the conservation law for heat conduction. A more general scalar conservation law for a free or moving boundary problem is to find $\{u,\Gamma\}$ such that

$$\nabla_{\underset{\sim}{X}}.\underset{\sim}{q}(u) = 0, \quad \underset{\sim}{X} \in \mathcal{D}^+ \cup \mathcal{D}^- \tag{6.1a}$$

with

$$u = 0, \quad [\underset{\sim}{q}.\underset{\sim}{n}]_-^+ = 0 \text{ on } \Gamma = \partial\mathcal{D}^+ \cap \partial\mathcal{D}^- \tag{6.1b,c}$$

and

$$u > 0 \text{ in } \mathcal{D}^+, \quad u < 0 \text{ in } \mathcal{D}^- \tag{6.1d}$$

in the fixed domain $\mathcal{D} = \mathcal{D}^+ \cup \Gamma \cup \mathcal{D}^-$ in which we use the same notation for row and column vectors. Here $\underset{\sim}{q}$ is a function of u and its derivatives which has a discontinuity at u = 0 of a type which enables (6.1a) to be interpreted in the sense of distributions. Also $\mathcal{D}$ may be a space-time domain with $\underset{\sim}{X} = (\underset{\sim}{x},t)$ in which case $\nabla_{\underset{\sim}{X}} = (\nabla_{\underset{\sim}{x}}, \frac{\partial}{\partial t})$. The Rankine-Hugoniot condition (6.1c) is automatically satisfied by solutions of the weak formulation

$$\int_{\mathcal{D}} (\underset{\sim}{q}.\nabla\phi)\,d\underset{\sim}{X} = 0 \text{ for all } \phi \in C_0^\infty(\mathcal{D}) \tag{6.2}$$

or equivalently, $\nabla_{\underset{\sim}{X}}.\underset{\sim}{q} = 0$ in the sense of distributions, as long as these solutions have the property that the regions $\mathcal{D}^+$ and $\mathcal{D}^-$ actually exist. A one-phase problem when written in this way would have a region $\mathcal{D}^0$ in which $u \equiv 0$ in place of $\mathcal{D}^+$ or $\mathcal{D}^-$.

For the problems considered in Chapter III

$$\mathcal{D} = \Omega \times (0,T), \quad \underset{\sim}{q}(u) = (-\nabla_{\underset{\sim}{x}} u, H), \quad H \in \beta(u) \tag{6.3}$$

where β is defined by (3.2) or (3.17), but other problems in this book may not be written so unambiguously in the form (6.1) as we shall now see for the dam problem.

Referring to Fig. 9, p. 58 , we let $\mathcal{D}$ be the region OA'B'D and recall that with the normalisation that $\rho g = 1$, the free boundary conditions are

$$p = 0, \quad [(\nabla p + (0,H)).\underset{\sim}{n}]_-^+ = 0, \text{ where } H(y) = \begin{cases} 1, & y < h(x) \\ 0, & y > h(x). \end{cases} \tag{6.4}$$

This suggests writing $-\underset{\sim}{q} = (\frac{\partial p}{\partial x}, \frac{\partial p}{\partial y} + H)$ in (6.1), so that

$$\frac{\partial H}{\partial y} + \nabla^2 p = 0 \text{ in } \mathcal{D} \tag{6.5a}$$

in the sense of distributions, where we now define

$$H(x,y) \in \beta(p(x,y)) \text{ a.e. in } \mathcal{D} \tag{6.5b}$$

with

$$\beta(p) = \begin{cases} 1 \ , & p > 0 \\ \in [0,1], & p = 0 \\ 0 \ , & p < 0 \end{cases} . \tag{6.5c}$$

The weak formulation of this problem is to find $\{p,H\}$ such that $p \in H^1(\mathcal{D})$, $p \geq 0$ a.e. in $\mathcal{D}$ and is prescribed on OABCD, $H \in L^\infty(\mathcal{D})$, (6.5b) holds and

$$\int_D \{\nabla p.\nabla\phi + H \frac{\partial \phi}{\partial y}\} \, dxdy = 0 \text{ for all } \phi \in \{H^1(\mathcal{D}) : \phi = 0 \text{ on OABCD}\}. \tag{6.6}$$

However, an alternative weak formulation is obtained by writing

$$\tilde{\underset{\sim}{q}} = \begin{cases} -\nabla p \ , & y < h(x) \\ (0,1), & y > h(x) \end{cases} \tag{6.6a}$$

so that

$$[\tilde{\underset{\sim}{q}}.\underset{\sim}{n}]_-^+ = 0 \tag{6.6b}$$

at the free boundary. If we define

$$u = \begin{cases} p \ , & y < h(x) \\ h(x) - y, & y > h(x) \end{cases} \tag{6.6c}$$

then $\underset{\sim}{\tilde{q}} = -(\frac{\partial u^+}{\partial x}, \frac{\partial u}{\partial y})$ in the usual notation, and we again have an example of (6.1) with

$$-\underset{\sim}{\nabla}.\underset{\sim}{\tilde{q}} = \frac{\partial^2 u^+}{\partial x^2} + \frac{\partial^2 u}{\partial y^2} = 0 \qquad (6.7)$$

in the sense of distributions. However, unlike p in (6.5), u^+ only is prescribed on the vertical boundaries and u satisfies the Neumann condition $\partial u/\partial y = -1$ on the top of the dam A'B'.

Thus there are at least two ways of formulating the rectangular dam problem as a conservation law (6.1) and as we mentioned on p. 40 , in general such formulations would not be equivalent for hyperbolic conservation laws such as (1.2). Here we have an elliptic field equation and we may conjecture that (6.5) and (6.7) are equivalent, although this has not been proved. However, we shall see that despite the ellipticity, extra conditions may be required to ensure the uniqueness of the weak solution for more general dam problems.

First let us make a few remarks about existence.

Theorem VI.1. [5, 36]

The weak formulation (6.5) has a solution for $\{p,H\}$.

Proof.

As in (3.13) we consider a family of nonlinear problems obtained by smoothing β, say by considering either the weak or classical formulations of

$$\frac{\partial H_\varepsilon}{\partial y} + \nabla^2 p_\varepsilon = 0 \text{ in } \mathcal{D}, \text{ where } H_\varepsilon(p_\varepsilon) = \begin{cases} 1\,, & p_\varepsilon > \varepsilon \\ p_\varepsilon/\varepsilon\,, & 0 \le p_\varepsilon \le \varepsilon. \\ 0\,, & p_\varepsilon < 0 \end{cases} \qquad (6.8)$$

Using the Schauder fixed point theorem, it can be proved that there is a unique solution to (6.8) and that $\|p_\varepsilon\|_{H^1}$ is bounded independently of ε. Then the usual compactness arguments give the existence of a solution of (6.5) as a limit as $\varepsilon \to 0$.

Existence has also been proved for (6.7), even for fairly general dams, [243, 244]. Also direct finite difference discretisations of (6.7) have been solved, [100], and indirect finite element approximations have been considered in [196].

We have seen after (2.65b) that the physical requirement that there is no inflow across the seepage face BC is automatically satisfied for the rectangular dam. However, for a sloping-sided dam, i.e. one with "batter", (Fig. 39), $\partial y/\partial n \geq 0$ on BC so that even though $\partial p/\partial n \leq 0$ on BC there may be physically unacceptable solutions of the free boundary problem (2.64, 2.65) unless we impose the inflow condition $\partial(p+y)/\partial n \leq 0$ on the seepage face. This is analogous to the examples on p. 94 and p. 159, and, for uniqueness, we must at least replace (6.9) by the variational inequality, [4, 5, 6, 36, 49]

$$\int_{\mathcal{D}} \{\nabla p.\nabla\phi + H \frac{\partial\phi}{\partial y}\}\, dxdy \geq 0 \tag{6.9a}$$

for all $\phi \in \{H^1(\mathcal{D}) : \phi = 0 \text{ on } OA \cup CD \text{ and } \phi \leq 0 \text{ on the dam boundary } ABC\}$.

(6.9b)

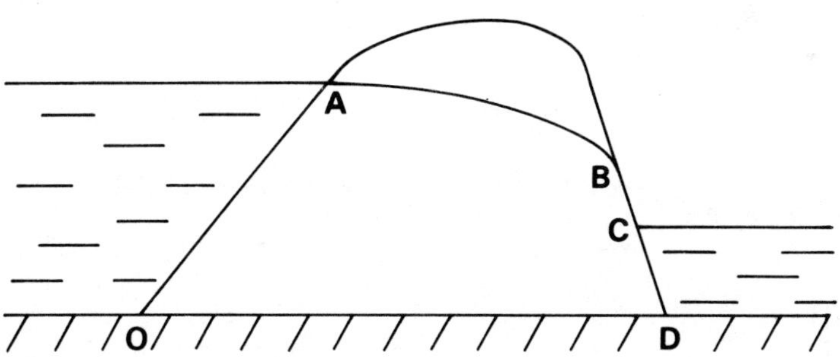

Fig. 39. Dam with Batter

Indeed, it can be shown that the solution of the regularised problem (6.8) converges to that of (6.9) as $\varepsilon \to 0$, [36]. However, the question of uniqueness is a topic of current research, [49] .

We conclude this section by mentioning a simple class of conservation laws (6.1) which can be used to solve some other commonly occurring free and moving boundary problems. Suppose the equation

$$\frac{\partial H}{\partial t} + \underset{\sim}{v}.\nabla H = \nabla^2 u + f, \quad H \in \beta(u) \tag{6.10}$$

holds in the sense of distributions in $\mathcal{D} = \Omega \times (0,T)$, where β is defined by (3.2) and $\underset{\sim}{v}$ is an arbitrary vector satisfying $\nabla.\underset{\sim}{v} = 0$. Thus, in (6.1a)

$$\underset{\sim}{q}(u) = (\nabla u - H\underset{\sim}{v}, -H) \tag{6.11}$$

and (6.1c) becomes, across the moving boundary $S(\underset{\sim}{x},t) = 0$

$$[H]_-^+ \frac{\partial S}{\partial t} = [\nabla u - H\underset{\sim}{v}]_-^+.\nabla S. \tag{6.12}$$

We thus have a fixed domain formulation for Stefan problems in bodies moving with a prescribed velocity $\underset{\sim}{v}$, such as occur in welding, [151] and continuous casting, [29]. We may also consider more general flow in the liquid region as long as the flow equations can be solved for $\underset{\sim}{v}$. Finally, taking β to be defined by (3.17), and $\underset{\sim}{v}$ to be parallel to the gravity field, we have a weak formulation of the unsteady dam problem and related Hele-Shaw problems [120] (p. 25). For travelling wave solutions of the Hele-Shaw problems, (6.12) reduces to (6.5), as expected from (1.39).

2 THE DERIVATION OF VARIATIONAL INEQUALITIES FROM A CLASS OF CONSERVATION LAWS

Using the notation of (6.1) and (6.10), consider the one-phase unknown boundary problem

$$\underset{\sim}{v}.\nabla_{\underset{\sim}{x}} H = L_{\underset{\sim}{x}} u \tag{6.13a}$$

where $\underset{\sim}{v}$ is a constant vector, $L_{\underset{\sim}{x}}$ a constant coefficient linear second order elliptic differential operator and H(u) has a finite jump discontinuity at $u = 0$. (6.13a) holds in the sense of distributions on $\mathcal{D} = \mathcal{D}^+ \cup \Gamma \cup \mathcal{D}^0$ where

$$\mathcal{D}^+ = \{\underset{\sim}{x} \in \mathcal{D}: u > 0\}, \quad \mathcal{D}^0 = \{\underset{\sim}{x} \in \mathcal{D}\backslash\Gamma: u = 0\}. \tag{6.13b}$$

The precise form of H is irrelevant for this section, so we will just consider

$$H = \begin{cases} \alpha u + L \text{ in } \mathcal{D}^+ \\ 0 \quad \text{in } \mathcal{D}^0 \end{cases} \tag{6.13c}$$

where $\alpha \geq 0$, $L > 0$.

Thus when the highest derivative (principal part) in $L_{\underset{\sim}{x}}$ is $\nabla^2_{\underset{\sim}{x}}$, the Rankine-Hugoniot conditions at Γ are

$$u = 0, \qquad \frac{\partial u}{\partial n_{\underset{\sim}{x}}} = L\underset{\sim}{v}.\underset{\sim}{n}_{\underset{\sim}{x}}. \tag{6.13d}$$

We now assume

(i) that $\partial\mathcal{D}$ is such that there is a subset $\partial\mathcal{D}_0 \subset \partial\mathcal{D}$ for which any $\underset{\sim}{x} \in \mathcal{D}$ can be written uniquely as

$$\underset{\sim}{x} = \underset{\sim}{x}_0 + s_0\,\underset{\sim}{v}, \quad \underset{\sim}{x}_0 \in \partial\mathcal{D}_0 \tag{6.14a}$$

where $\underset{\sim}{x}_0 + s\underset{\sim}{v} \in \mathcal{D}$ for all $s \in (0,s_0]$ (6.14b)

and

(ii) that Γ is such that any line parallel to $\underset{\sim}{v}$ meets it just once. We note that we may be able to apply a maximum principle in $\mathcal{D}^+$ to deduce that $\left.\frac{\partial u}{\partial n_{\underset{\sim}{x}}}\right|_\Gamma < 0$ and hence, from (6.13d) that $\underset{\sim}{v}.\underset{\sim}{n}_{\underset{\sim}{x}} < 0$ which prohibits lines parallel to $\underset{\sim}{v}$ meeting Γ more than once. Thus, as in Fig. 40, there is a unique s* such that

$$\underset{\sim}{x}_0 + s\underset{\sim}{v} \in \begin{cases} \mathcal{D}^0, & s \in (0,s^*) \\ \mathcal{D}^+, & s > s^* \end{cases} \tag{6.15}$$

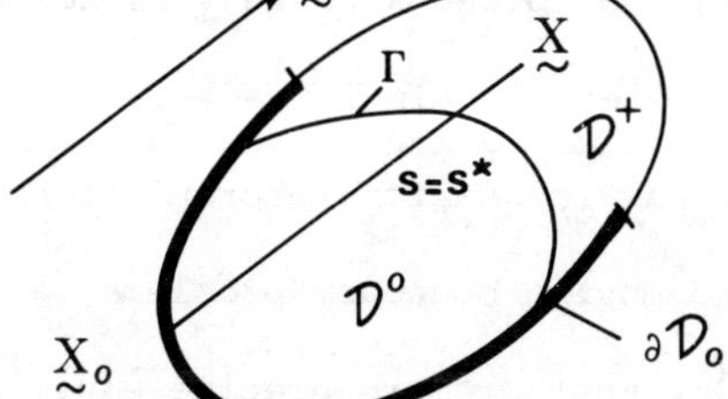

Fig. 40

These assumptions mean we can integrate u in the $\underset{\sim}{v}$ direction to obtain a new dependent variable which has a continuous gradient at Γ. Putting

$$U(\underset{\sim}{X}) = \int_0^{s_0} u(\underset{\sim}{X}_0 + s\underset{\sim}{v})\, ds \tag{6.16}$$

we see from (6.15) that

$$U = 0 \text{ in } \mathcal{D}^0, \quad U > 0 \text{ in } \mathcal{D}' \tag{6.17a}$$

and

$$\underset{\sim}{v}.\nabla_{\underset{\sim}{X}} U = u. \tag{6.17b}$$

Hence (6.13a) becomes

$$\underset{\sim}{v}.\nabla_{\underset{\sim}{X}} H = L_{\underset{\sim}{X}}(\underset{\sim}{v}.\nabla_{\underset{\sim}{X}} U) \tag{6.18}$$

which is

$$\underset{\sim}{v}.\nabla_{\underset{\sim}{X}}(H - L_{\underset{\sim}{X}} U) = 0 \tag{6.19}$$

since, in view of our assumptions, $L_{\underset{\sim}{X}}$ and $\underset{\sim}{v}.\nabla_{\underset{\sim}{X}}$ commute, and, on Γ

$$U = \underset{\sim}{v}.\nabla_{\underset{\sim}{X}} U = 0. \tag{6.20}$$

Moreover, (6.19) implies that $H - L_{\underset{\sim}{X}} U = F(\underset{\sim}{X}_0)$ and $F(\underset{\sim}{X}_0) = 0$ if $\underset{\sim}{X}_0 \in \partial\mathcal{D}^0$. Hence, since $H = \alpha\underset{\sim}{v}.\nabla_{\underset{\sim}{X}} U + L$ in $\mathcal{D}^+$, U satisfies the linear complementarity problem

$$U \geq 0, \quad \alpha\underset{\sim}{v}.\nabla_{\underset{\sim}{X}} U - L_{\underset{\sim}{X}} U + L - F(\underset{\sim}{X}_0) \geq 0 \tag{6.21a}$$

$$U(\alpha\underset{\sim}{v}.\nabla_{\underset{\sim}{X}} - L_{\underset{\sim}{X}} U + L - F(\underset{\sim}{X}_0)) = 0. \tag{6.21b}$$

Hence we can obtain a variational inequality of the form described in Chapter IV as long as we can find $F(\underset{\sim}{X}_0) = (H - L_{\underset{\sim}{X}} U)(\underset{\sim}{X}_0)$.

This approach to deriving variational inequalities was first proposed in [15 , 19], and the parabolic case was studied in [119]. Unfortunately the theory does not readily generalise to cases where $\underset{\sim}{v}$ is non-constant or where $L_{\underset{\sim}{X}}$ has variable coefficients, except in simple cases.

We have thus found variational inequality reformulations of some members of the class (6.1) of free and moving boundary problems, all which have weak formulations. We will consider another more precise equivalence between weak and variational formulations for the one-phase Stefan problem at the end of this section, but let us first give some examples to illustrate the analysis above.

(i) *One-phase Stefan and Hele-Shaw Problems*

Setting $\mathcal{D} = \Omega \times (0,T)$, $\underset{\sim}{X} = (\underset{\sim}{x}, t)$, $L_{\underset{\sim}{X}} = \nabla^2_{\underset{\sim}{x}}$, $\underset{\sim}{v} = (0,1)$ and $\partial\mathcal{D}_0 = \{(\underset{\sim}{x},0) : \underset{\sim}{x} \in \Omega\}$, we can write $\underset{\sim}{X} = (\underset{\sim}{x},0) + t\underset{\sim}{v}$, which is (6.14a), and (6.16) is

$$U = \int_0^t u(\underset{\sim}{x},\tau)\, d\tau, \tag{6.22}$$

which is just (4.84). (6.21) reduces to (4.90).

(ii) *An electrochemical machining problem*

To solve (1.32, 1.39) with a fixed cathode $y = c(x)$, Fig. 41, we set ($\phi = -p = u$) $\underset{\sim}{X} = (x,y)$, $\mathcal{D} = \{(x,y): 0 < x < a,\ c(x) < y < b\}$, $L_{\underset{\sim}{X}} = \nabla^2_{\underset{\sim}{X}}$, $\underset{\sim}{v} = (0,-1)$, $\alpha = 0$, $L = 1$ and $\partial\mathcal{D}_0 = \{(x,b) : 0 < x < a\}$.

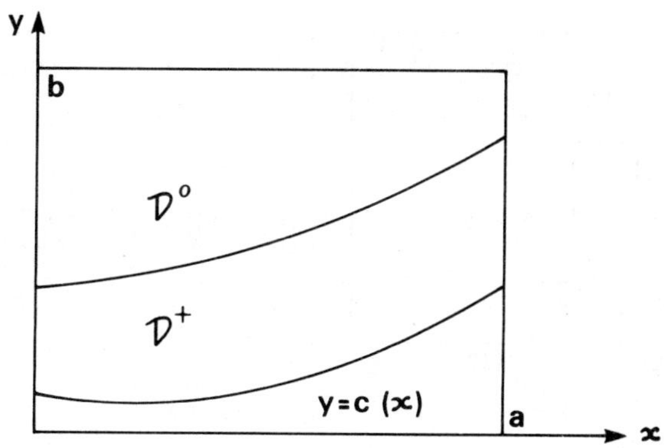

Fig. 41

Then (6.16) is

$$U = \int_y^b u(x,y')\,dy' \tag{6.23}$$

and (6.21) reduces to (4.90) with $\alpha = 0$. The boundary conditions on U are of Dirichlet or Neumann type except on the cathode where $\partial U/\partial y = -1$. This oblique derivative condition means that U satisfies a variational inequality for a non-symmetric bilinear form, [94].

(iii) *One-dimensional ablation*

Putting $k = \rho = L = 1$ in (5.4) to (5.7), and replacing u by $-u$,

$$\frac{\partial H}{\partial t} - Q\frac{\partial H}{\partial x} = \frac{\partial^2 u}{\partial x^2} - Q\frac{\partial u}{\partial x} \text{ in } \mathcal{D} = (0,d) \times (0,T) \tag{6.24}$$

where

$$H = \begin{cases} u+1, & u > 0 \\ 0, & u = 0 \end{cases}, \tag{6.25}$$

which is related to (5.8), but this is a different formulation from that used on p. 156.

Now by (5.4) and the maximum principle, $ds/dt \geq -Q$ so that lines parallel to $x + Qt = 0$ meet $x = s(t)$ just once (Fig. 42).

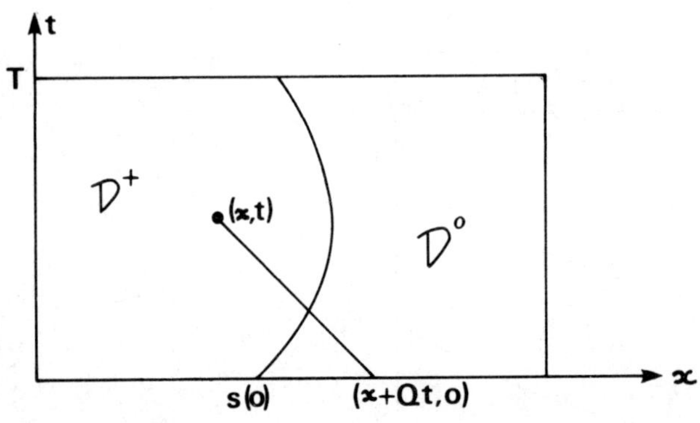

Fig. 42

Hence we set

$$\underset{\sim}{X} = (x,t), \quad L_{\underset{\sim}{X}} = \frac{\partial^2}{\partial x^2} - Q\frac{\partial}{\partial x},$$

$$\underset{\sim}{v} = (-Q,1), \quad \text{so that } \underset{\sim}{X} = (x+Qt,0) + t(-Q,1).$$

$\partial\mathcal{D}_0$ is $\{(x,0): 0 < x < d\}$ and (6.16) is

$$U = \int_0^t u(x + Q(t-\tau),\tau)\,d\tau . \tag{6.26}$$

If the initial condition is $u(x,0) = u_0 = 0$ in $x < s(0)$, then, in (6.21),

$$F(\underset{\sim}{X}_0) = \begin{cases} u_0(x+Qt) + 1, & u_0 > 0 \\ 0 \quad , & u_0 = 0 \end{cases} . \tag{6.27}$$

On $x = d$, $U = 0$ but on $x = 0$ we again have an oblique derivative boundary condition

$$\frac{\partial U}{\partial t} - Q\frac{\partial U}{\partial x} = u(0,t). \tag{6.28}$$

The transformation (6.26) was first derived in [236], (see also [156]).

(v) *Dam with Batter*

Referring to Fig. 39, we let OABCD be the curve $y = z(x)$, $0 < x < d$ and set $\underset{\sim}{X} = (x,y)$, $\underset{\sim}{v} = (0,-1)$, $\alpha = 0$, $L = 1$ and $L_{\underset{\sim}{x}} = \nabla^2_{\underset{\sim}{x}}$. $\partial\mathcal{D}_0$ is $\{(x,z(x)):0 < x < d\}$ and, writing $\tilde{y} = z(x)-y$, $(x,\tilde{y}) = (x,z(x)) + y(0,-1)$. (6.16) becomes

$$U(x,y) = \int_0^y p(x,z(x)-y')\,dy'$$

which is simply (2.66) after a change of variable. (6.21) becomes (2.67) but only in the region vertically below AB. When OA or BD are not vertical, a new approach becomes necessary and some progress has been made using the theory of so-called quasi-variational inequalities, [12, 14, 181]. However, to describe this theory here would take us beyond the scope of this book.

We conclude this section by noting an equivalence between two discretisations of the weak and variational formulations of the one phase Stefan problem. The explicit finite difference scheme (3.23) is, in a simple case

$$H_i^{n+1} = H_i^n + k\delta^2 u_i^n/h^2, \quad \delta^2 u_i^n = u_{i+1}^n + u_{i-1}^n - 2u_i^n, \tag{6.29a}$$

$$u_i^{n+1} = \begin{cases} H_i^{n+1} - L, & H_i^{n+1} > L \\ 0 \quad , & H_i^{n+1} \in [0,L] \\ H_i^{n+1} \quad , & H_i^{n+1} < 0 \end{cases} \tag{6.29b}$$

with initial and boundary conditions

$$H_i^0 = u_i^0 = 0, \quad u_m^n = 0; \quad u_0^n = u_0 = \text{const}, \; n > 0, \tag{6.29c}$$

say. The corresponding explicit scheme (4.103) is

$$U_i^{n+1} = \{U_i^n + \frac{k}{h^2}\delta^2 U_i^n - kL\}^+ \tag{6.30a}$$

for $1 \le i < m$, with

$$U_0^n = nku_0, \quad U_i^0 = 0, \quad U_m^n = 0. \tag{6.30b}$$

We now assert that the solutions of (6.29) and (6.30) are related by

$$k\, u_i^n = U_i^{n+1} - U_i^n . \tag{6.31}$$

To prove this, we first note that providing $k/h^2 \le \frac{1}{2}$, U_i^n, u_i^n satisfy the discrete maximum principles

$$u_0 \ge u_i^{n+1} \ge u_i^n \ge 0, \tag{6.32a}$$

$$nku_0 \ge U_i^n \ge 0, \quad U_i^{n+1} \ge U_i^n , \tag{6.32b}$$

$$\delta^2 U_i^n \ge 0, \quad U_{i+1}^n \ge U_i^n, \tag{6.32c}$$

for all $i \in [1,M)$. The second of (6.32b) is proved on p. 142, and the remainder may be proved in a similar way. Now for U_i^n satisfying (6.30), define

$$\tilde{u}_i^n = (U_i^{n+1} - U_i^n)/k \tag{6.33a}$$

and

$$\tilde{H}_i^n = \tilde{H}_i^{n-1} + \frac{1}{h^2}\delta^2(U_i^n - U_i^{n-1}), \quad n \ge 1 \tag{6.33b}$$

with $\tilde{H}_i^0 = 0$. All we have to show is that $\tilde{u}_i^n = (\tilde{H}_i - L)^+$. However, summing (6.33b) yields

$$\tilde{H}_i^n = \frac{1}{h^2}\,\delta^2\, U_i^n \geq 0$$

and if $\tilde{u}_i^n > 0$, from (6.32c) and (6.33a), $U_i^{n+1} > 0$ and so

$$\tilde{u}_i^n = \frac{1}{h^2}\,\delta^2 U_i^n - L = \tilde{H}_i^n - L.$$

Equally, if $\tilde{u}_i^n = 0$, then $U_i^{n+1} = U_i^n$ and from (6.30a) $\frac{1}{h^2}\,\delta^2 U_i^n - L = \tilde{H}_i^n - L \leq 0$ and our assertion is proved.

3. CONCLUSION

The last two sections have illustrated the greatest difficulty associated with weak and variational formulations of free and moving boundary problems and that is the uncertainty about whether any particular classical model is one of the few which is suitable for such a formulation. We have tried to indicate some rules for deciding about this suitability but we have only been able to classify conservation laws with appropriate discontinuities in the dependent variables as being amenable to a weak formulation. The narrowness of this classification is demonstrated by the Muskat problem (p. 160), which is certainly a conservation statement. As far as variational inequalities are concerned, we have said nothing about classes of suitable problems. Their most natural source is as obstacle problems in elasticity but most of the other problems we have analysed variationally have required preliminary integral transformations. Although these transformations lack any obvious physical interpretation, we have tried to group them mathematically in the last section.

The applications in this Chapter and in Chapter V also reveal other drawbacks and limitations. A weak or variational formulation may or may not lead to a fixed domain method of solution and even if it does, there may be non-uniqueness. We also observe that all the examples we have been able to treat in this book by fixed domain methods have the characteristic that the unknown boundary divides the fixed domain into two regions in which a certain function of the dependent variable is either positive, zero or negative.

Against all this uncertainty must be set the ease with which the numerical schemes described in Chapters III and IV can be implemented and justified. Indeed this is the raison d'être of this book. Theoretically, the existence uniqueness, convergence and numerical stability results we have described are far simpler then the few that are known for classical models, and we have not even mentioned some of the valuable information concerning properties of the unknown boundary which can be derived from variational inequality formulations, [140]. From the practical point of view, results such as the free boundary shape displayed on p. 125 or the mushy regions described on p. 167 could only have been obtained by classical methods of a very ad hoc nature.

References

1 Agrawal, H.C. Biot's Variational Principle for Moving Boundary Problems in [180], 242-250.

2 Aitchison, J.M. The Numerical Solution of Planar and Axisymmetic Flow. Rep. RL-78-035. Rutherford Lab. Harwell, Oxon, U.K.

3 Albrecht, J., Collatz, L., Hoffman, K-H., (eds.) Proceedings Oberwolfach Symposium (1980) Birkhauser-Verlag.

4 Alt, H.W. A Free Boundary Problem Associated with the Flow of Ground Water. Arch. Rat. Mech. Anal. 64, (1977), 111-126.

5 Alt, H.W. Strömungen Durch Inhomogene Poröse Medien mit Freiem Rand. J. Reine Ang. Math. 305 (1979), 89-115.

6 Alt, H.W. Numerical Solution of Steady State Porous Flow Free Boundary Problems. Numer. Math. 36 (1980), 73-98.

7 Andrews, J and Atthey, D.R. On the Motion of an Intensely Heated Evaporating Boundary. J. Inst. Maths. Applics. 15 (1975), 59-72.

8 Aronson, D.G. Regularity Properties of Flows through Porous Media. SIAM J. Appl. Math. 17 (1969), 461-467.

9 Atthey, D.R. A Finite Difference Scheme for Melting Problems. J. Inst. Maths. Applics. 13 (1974), 353-366.

10 Avdonin, N.A. The Mathematical Description of Crystallisation Processes (in Russian) Riga: Zinatne 1980.

11 Baiocchi, C. Su un problema di frontiera libera connesso a questioni di idraulica. Ann. Mat. Pura. Appl. XCII (1972), 107-127.

12 Baiocchi, C., Comincioli, V., Magenes, E., and Pozzi, G.A. Free boundary problems in the theory of fluid flow through porous media: existence and uniqueness theorems. Ann. Mat. Pura Appl. XCVII (1973), 1-82.

13 Baiocchi, C., Comincioli, V., Guerri, L. and Volpi, G. Free boundary problems in the theory of fluid flow through porous media: a numerical approach. Calcolo 10 (1973), 1-85.

14 Baiocchi, C. Studio di un problema quasi-variazionale connesso a problemi di frontiera libera. Boll. U.M.I. (4) 11 (1975), 589-613.

15 Baiocchi, C. Problèmes à frontiere libre et inéquations variationnelles. C.R. Acad. Sci. Paris 283 (1976), 29-32.

16 Baiocchi, C. and Pozzi, G. An evolution variational inequality related to a diffusion absorption problem. Appl. Math. Optim. 4 (1976), 304-314.

17 Baiocchi, C. and Pozzi, G. Error estimates and free boundary convergence for a finite difference discretisation of a parabolic variational inequality. RAIRO An. Num. 11 (1977), 315-340.

18 Baiocchi, C. Estimation d'erreur dans L^{∞} pour les inequations à obstacle. in Mathematical Aspects of Finite Element Methods, Rome, 1975. Springer Lect. Notes No. 606 (1977) 27-34.

19 Baiocchi, C. Variational inequalities and free boundary problems in [63], 25-33.

20 Bald, W.B. and Crowley, A.B. On defining the thermal history of cells during the freezing of biological materials. J. Microscopy 117 (1979), 395-409.

21 Barenblatt, G.I. Similarity, Self-Similarity and Intermediate Asymptotics. Consultants Bureau, New York & London, 1979.

22 Bear, J. Dynamics of Fluids in Porous Media. Envir. Sci. Sér. Elsevier, New York, 1972.

23 Benjamin, T.B. and Olver, P. Hamiltonian Structure, Symmetries and Conservation Laws for Water Waves. Preprint 1981.

24 Bensoussan, A. and Lions, J.L. Applications des inéquations variationelles en contrôles stochastique. Dunod, Paris (1978).

25 Berger, A.E., Ciment, M. and Rogers, J.C.W. Numerical Solution of a Diffusion Consumption Problem with a Free Boundary. SIAM J. Num. Anal. 16 (1975) 646.

26 Berger, A. and Falk, R. An error estimate for the truncation method for the solution of parabolic variational inequalities. Math. Comp. 31 (1977), 619-628.

27 Bermudez, A. and Saguez, C. Etude Numerique d'un Problème de Solidification d'un Alliage. Preprint 1981.

28 Birkhoff, G. and Zarantonello, E.H. Jets, Wakes and Cavities. Academic Press, New York, 1957.

29 Blackwell, J. and Ockendon, J.R. Exact Solution of a Stefan Problem Relevant to Continuous Casting. Int. J. Heat Mass Transfer (in press).

30 Boley, B.A. The Embedding Technique in Melting and Solidification Problems. in [180], 150-172.

31 Boley, B.A. An Applied Overview of Moving Boundary Problems. in [249], 205-232.

32 Bossavit, A. and Damlamian, A. Homogenization of the Stefan Problem and Application to Magnetic Composite Media. IMA J. Appl. Math. 27 (1981), 319-334.

33 Brezis, H. On some degenerate nonlinear parabolic equations. in Symposia in Pure Math. 18 Amer. Math. Soc., Providence R.I. (1970), 28-38.

34 Brezis, H. Problèmes Unilateraux. J. Math. Pures et App. 51 (1972), 1-168.

35 Brezis, H. Operateurs Maximaux Monotones. North-Holland Mathematical Studies 5. North Holland, Amsterdam, 1973.

36 Brezis, H., Kinderlehrer, D. and Stampacchia, G. Sur un nouvelle formulation du problème d'écoulement à travers une digne. C.R. Acad. Sc. Paris 287 (1978), 711-714.

37 Brezis, H. and Sibony, M. Equivalence de deux inequations variationelles et applications. Arch. Rat. Mech. Anal. 41 (1971), 254-265.

38 Brezis, H. and Stampacchia, G. Sur la régularité de la solution d'inéquations elliptiques. Bull. Soc. Math. Fr. 96 (1968), 153-180.

39 Brezis, H. and Stampacchia, G. The hodograph method in fluid dynamics in the light of variational inequalities. Arch. Rat. Mech. Anal. 61 (1976), 1-18.

40 Brezzi, F., Hager, W.W. and Raviart, P.A. Error estimates for the finite element solution of variational inequalities. Numer. Math. 28 (1977), 431-443.

41 Brezzi, F. and Sacchi, G. A finite element approximation of variational inequalities related to hydraulics. Calcolo 13, (1976), 259-273.

42 Buckmaster, J. Viscous Sheets Advancing over Dry Beds. J. Fluid Mech. 81 (1977), 735-756.

43 Buckmaster, J. A Mathematical Description of Open and Closed Flame Tips. Comb. Sci. Tech. 20 (1979) 33-40.

44 Burridge, R. and Keller, J.B. Peeling, Slipping and Cracking - Some One-Dimensional Free-Boundary Problems in Mechanics. SIAM Rev., 20 (1978), 31-61.

45 Cameron, A. Basic Lubrication Theory. Halsted Press, Wiley, New York, 1976.

46 Cannon, J.R. and Hill, C.D. Existence, Uniqueness, Stability and Monotone Dependence in a Stefan Problem for the Heat Equation. J. Math. Mech. 17 (1967), 1.

47 Capriz, G. and Cimatti, G. On Some Singular Perturbation Problems in the Theory of Lubrication. Appl. Math. Optim. 4 (1978), 285-297.

48 Carré, B.A. The Determination of the Optimum Accelerating Factor for Successive Over-relaxation. Comp. J. 4 (1961), 73-78.

49 Carrillo Menendez, J. and Chipot, M. Sur l'unicité de la solution du problème de l'ecoulement à travers une digne. C.R. Acad. Sc. Paris 292 (1981), 191-194.

50 Carslaw, H.S. and Jaeger, J.C. Conduction of Heat in Solids. Oxford Univ. Press, Oxford, 1959.

51 Cea, J., Glowinski, R. and Nedelec, J.C. Application des méthodes optimisation, de différences et d'éléments finis à l'analyse numérique de la torsion élasto-plastique d'une barre cylindrique. Cahier de l'INRIA No. 12 (1974), 7-138.

52 Chadam, J. and Ortoleva, P. The Stabilizing Effect of Surface Tension on the Development of Free Boundaries. in [105].

53 Chandrasekhar, S. Hydrodynamic and Hydromagnetic Stability. Oxford Univ. Press, Oxford, 1961.

54 Chen, B. and Saffman, P.G. Numerical Evidence for the Existence of New Types of Gravity Waves of Permanent Form in Deep Water. Stud. Appl. Math. 62 (1980), 1-21.

55 Christopherson, D.G. A New Mathematical Method for the Solution of Film Lubrication Problems. Proc. Inst. Mech. Eng. 146 (1941), 126-135.

56 Christopherson, D.G. and Southwell, R.V. Relaxation Methods Applied to Engineering Problems III. Problems Involving

Two Independent Variables. Proc. Roy. Soc. A 168 (1938), 317-350.

57 Ciarlet, P.G. The Finite Element Method for Elliptic Problems. North Holland, Amsterdam, 1978.

58 Ciavaldini, J.F. Analyse Numérique d'un Problème de Stefan à Deux Phases Par un Methode d'Elements Finis. SIAM J. Numer. Anal. 12 (1975), 464-487.

59 Cimatti, G. On a Problem of the Theory of Lubrication Governed by a Variational Inequality. Appl. Math. Optim. 3 (1977), 227-242.

60 Comincioli, V., Guerri, L. and Volpi, G. Analisi numerica di un problema di frontiera libera connesso col moto di un fluido attraverso un mezzo poroso. Pub. N17 (1971) Lab. Anal. Num., C.N.R., Pavia.

61 Constable, T.B. and Evans, N.T.S. A Method for Measuring the Spatial Distribution of Tissue Oxygen Removal Rates. Resp. Physiol., 25, (1975), 175-190.

62 Cottle, R.W. Numerical Methods for Complementarity Problems in Engineering and Applied Science. Proc. Comp. Meth. in Appl. Sci. Eng. Springer Lec. Notes No. 704 (1979), 37-52.

63 Cottle, R., Giannessi, F. and Lions, J.L. (eds.) Variational Inequalities and Complementarity Problems. Wiley, New York, 1980.

64 Cottle, R.W., Golub, G.H. and Sacher, R.S. On the Solution of Large, Structured Linear Complementarity Problems: the Block Partitioned Case. Appl. Math. Optim. 4 (1978), 347-363.

65 Cottle, R.W. and Sacher, R.S. On the Solution of Large Structured Linear Complementarity Problems: the Tridiagonal Case. Appl. Math. Optim. 3 (1977), 321-340.

66 Couch, E.J., Keller, H.H. and Watts, J.W. Permafrost Thawing Around Producing Oil Wells. J. Can. Pet. Tech., April-June 1970, 107-111.

67 Courant, R. and Friedrichs, K.O. Supersonic Flow and Shock Waves. Interscience, New York, 1948.

68 Coyne, J.C. and Elrod, H.G. Conditions for the Rupture of a Lubricating Film, Pt. 1. Theoretical Model. J. Lub. Tech. 92 (1970), 451-456.

69 Crank, J. Finite-difference Methods. in [180], 192-207.

70 Crank, J. The Numerical Solution of Free and Moving Boundary Problems. Oxford Univ. Press, Oxford (in press).

71 Crank, J. and Crowley, A.B. Isotherm Migration Along Orthogonal Lines in Two Dimensions. Int. J. Heat Mass Transf. 21 (1978), 393-398.

72 Crank, J. and Gupta, R.S. A Moving Boundary Problem Arising from the Diffusion of Oxygen in Absorbing Tissue. J. Inst. Maths. Applics. 10 (1972), 19-33.

73 Crank, J. and Phale, R.D. Melting Ice by the Isotherm Migration Method. Bull. Inst. Maths. Applics. 9 (1973), 12-14.

74 Crowley, A.B. On the Weak Solution of Moving Boundary Problems. J. Inst. Maths. Applics. 24 (1979), 43-57.

75 Crowley, A.B. Flame Fronts, Crystal Growth and Weak Solutions. in [3].

76 Crowley, A.B. and Ockendon, J.R. A Stefan Problem with a Non-monotone Boundary. J. Inst. Maths. Applics. 20 (1977), 269-281.

77 Crowley, A.B. and Ockendon, J.R. On the Numerical Solution of an Alloy Solidification Problem. Int. J. Heat Mass. Transf. 22 (1979), 941-947.

78 Cryer, C.W. The Solution of a Quadratic Programming Problem Using Systematic Overrelaxation. SIAM J. Control 9 (1971), 385-392.

79 Cryer, C.W. The Method of Christopherson for Solving Free Boundary Problems for Infinite Journal Bearings by Means of Finite Differences. Math. Comp. 25, (1971) 435-443.

80 Cryer, C.W. A Survey of Steady-State Porous Flow Free Boundary Problems. M.R.C. Rep. No. 1657 (1976), Math. Res. Cent., Univ. of Wisconsin.

81 Cryer, C.W. A Survey of Trial Free Boundary Methods for the Numerical Solution of Free Boundary Problems. M.R.C. Rep. No. 1693 (1977), Math. Res. Cent., Univ. of Wisconsin.

82 Cryer, C.W. A Bibliography of Free Boundary Problems. M.R.C., Rep. No. 1793 (1977), Math. Res. Cent., Univ. of Wisconsin.

83 Cryer, C.W. Successive Overrelaxation Methods for Solving Linear Complementarity Problems Arising from Free Boundary Problems. in [157], Vol. 1, 109-132.

84 Damlamian, A. Some Results on the Multiphase Stefan Problem. Comm. Part. Diff. Eqns., 2 (1977), 1017-1044.

85 Dobrovol'skaya, Z.N. On Some Problems of Similarity Flow of Fluid with a Free Surface. J. Fluid Mech. 36 (1969), 805-829.

86 Donnelly, J.D.P. A Model for Non-Equilibrium Thermodynamic Process Involving Phase Changes. J. Inst. Maths. Applics. 24 (1979), 425-438.

87 Dowson, D. and Taylor, C.M. Cavitation in Bearings. Ann. Rev. Fluid Mech. 11 (1979), 35-66.

88 Drake, L.D. and Shreve, R.L. Pressure Melting and Regelation of Ice by Round Wires. Proc. R. Soc. Lond. A 332 (1973), 51-83.

89 Duvaut, G. Résolution d'un problème de Stéfan. C.R. Acad. Sc. Paris 276 (1973), 1461-1463.

90 Duvaut, G. in [105].

91 Duvaut, G. and Lions, L. Inequalities in Mechanics and Physics. Springer, Berlin, 1976.

92 Elliott, C.M. On a Variational Inequality Formulation of an Electrochemical Machining Moving Boundary Problem and its Approximation by the Finite Element Method. J. Inst. Maths. Applics. 25, (1980), 121-131.

93 Elliott, C.M. On the Finite Element Approximation of an Elliptic Variational Inequality Arising from an Implicit Time Discretization of the Stefan Problem. I.M.A. J. Num. Anal. 1 (1981), 115-125.

94 Elliott, C.M. A Variational Inequality Formulation of an Electrochemical Machining Free Boundary Problem. in [105].

95 Elliott, C.M. A Class of Ordinary Differential Equations with a Discontinuous Nonlinearity and the Backward Euler Rule. (1981) preprint.

96 Elliott, C.M. and Janovsky, V. A Finite Element Discretisation of a Variational Inequality Formulation of a Hele-Shaw Moving Boundary Problem. Proc. MAFELAP 1978, Whiteman, J.R. (ed.) Academic Press, London, 1979.

97 Elliott, C.M. and Janovsky, V. A Variational Inequality Approach to Hele-Shaw Flow with a Moving Boundary. Proc. Roy. Soc. Edin. 88A (1981), 93-107.

98 Elliott, C.M. and Janovsky, V. An error-estimate for a Finite Element Approximation of an Elliptic Variational Inequality Formulation of a Hele-Shaw Moving Boundary Problem. (1981) preprint.

99 Elliott, C.M. and McKee, S. On the Numerical Solution of an Integro-differential Equation Arising from Wave-power Hydraulics. B.I.T. (1981) (in press).

100 Elliott, C.M. and Mason, C.J. A Weak Solution Method for a Class of Free Boundary Problems. in [3].

101 Eyres, N.R., Hartree, D.R., Ingham, J., Jackson, R., Sarjant, R.J., and Wagstaff, J.B. The Calculation of Variable Heat Flow in Solids. Phil. Trans. Roy. Soc. Lond. A 240 (1946), 1-57.

102 Falk, R.S. Error Estimates for the Approximation of a Class of Variational Inequalities. Math. Comp. 28, (1974), 963-971.

103 Fasano, A. and Primicerio, M. Convergence of Huber's Method for Heat Conduction Problems with Change of Phase. Z. Angew. Math. Mech. 53 (1973), 341-348.

104 Fasano, A., Primicerio, M. and Rubinstein, L. A Model Problem for Heat Conduction with a Free Boundary in a Concentrated Capacity. J. Inst. Maths. Applics. 26 (1980), 327-348.

105 Fasano, A. and Primicerio, M. (eds.) Proceedings Montecatini Symposium on Free and Moving Boundary Problems (1981), Springer (in press).

106 Fichera, G. Problemi Elastostatii con Vincoli Unilaterati. Atti Accad. Naz. Lincei Mem. Cl. Sci. Fiz. Mat. Natur. Sez. Ia (8) VII (1964), 91-140.

107 Fichera, G. Boundary Value Problems of Elasticity with Unilateral Constraints. Handbuch der Physik VIa/2, Mechanics of Solids II Springer (1972), 391-424.

108 Filippov, A.F. Differential Equations with Discontinuous Right-Hand Side. Amer. Math. Soc. Transl. 42 (1960), 199-231.

109 Fix, G. Numerical Methods for Alloy Solidification. in [249], 109-128.

110 Forsyth, P. and Rasmussen, H. Comparison of Variational Inequality and Front Tracking Solution of an Electrochemical Machining Problem. Utilas Mathem. 17 (1980), 3-15.

111 Fox, L. The Stefan Problem: Moving Boundaries in Parabolic Equations. in A Survey of Numerical Methods for Partial Differential Equations, Gladwell, I. and Wait, R. (eds.). Oxford Univ. Press, Oxford, 1979.

112 Fraenkel, L.E. and Berger, M. A Global Theory of Steady Vortex Rings in an Ideal Fluid. Acta Math. 132 (1974), 13-51.

113 Frehse, J. Two-dimensional Variational Problems with Thin Obstacles. Math. Z. 143 (1975), 279-288.

114 Friedman, A. The Stefan Problem in Several Space Variables. Trans. Am. Math. Soc. 133 (1968), 51-87.

115 Friedman, A. Parabolic Variational Inequalities in One Space Dimension and Smoothness of the Free Boundary. J. Funct. Anal. 18 (1975), 151-176.

116 Friedman, A. and Kinderlehrer, D. A One Phase Stefan Problem. Indiana Univ. Math. J. 24, (1975), 2005-2035.

117 Furzeland, R. A Comparative Study of Numerical Methods for Moving Boundary Problems. J. Inst. Maths. Applics. 26 (1980), 411-429.

118 Garabedian, P.R. Free Boundary Flows of a Viscous Liquid. Comm. Pure Appl. Math. 19 (1966), 421-434.

119 Gastaldi, F. About the Possibility of Setting Stefan-like Problems in Variational Form. Boll. Un. Mat. Ital. 16-A (1979), 148-156.

120 Gilardi, G. The Evolution Dam Problem. in [157], Vol. 1, 209-218.

121 Glowinski, R. Numerical Methods for Nonlinear Variational Problems. Lecture Notes Tata Institute, Bangalore (1979).

122 Glowinski, R. and Lanchon, H. Torsion Élastoplastique d'une Barre Cylindrique de Section Multiconnexe. J. Mecanique, 12 (1973), 151-171.

123 Glowinski, R., Lions, J.L. and Tremolieres, R. Analyse Numerique des Inequations (2 vols.). Dunod, Paris, 1976.

124 Graveleau, J.L. and Jamet, P. A Finite Difference Approach to Some Degenerate Nonlinear Parabolic Equations. SIAM J. Appl. Math. 20 (1971), 199-223.

125 Grinberg, G.A. and Chekmareva, O.M. Motion of the Phase Interface in the Stefan Problem. Sov. Phys. Tech. Phys. 15 (1971), 1579-1583.

126 Gustafsson, B. Applications of Variational Inequalities to a Moving Boundary Problem for Hele-Shaw Flows. Trita-Mat-1981-9, Dept. of Maths., Royal Inst. Tech., Stockholm, 1981.

127 Hájek, O. Discontinuous Differential Equations I and II. J. Diff. Eqns. 32 (1979), 149-170, 171-185.

128 Hansen, E. and Hougaard, P. On a Moving Boundary Problem from Biomechanics. J. Inst. Maths. Appl. 13 (1974), 385-398.

129 Hougard, P. Some Solutions of a Free Boundary Problem Related to Electrochemical Machining. Rep. No. 59, Dec. 1977, Tech. Univ. Denmark.

130 Huber, A. Hauptaufsätze uber das Fortschreitten der Schmelzgrenze in einem Linearen Leiter. Z. Ang. Math. Mech. 49 (1939) 1-21.

131 Isaacson, E. and Keller, H.B. Analysis of Numerical Methods. Wiley, New York, 1966.

132 Jeffrey, A. and Taniuti, T. Nonlinear Wave Propagation with Applications to Physics and Magnetohydrodynamics. Academic Press, New York, 1964.

133 Jerome, J.W. Nonlinear Equations of Evolution and a Generalized Stefan Problem. J. Diff. Eqns. 26, (1977), 240-261.

134 Johnson, C. A Convergence Estimate for an Approximation of a Parabolic Variational Inequality. SIAM J. Numer. Anal. 13 (1976), 599-606.

135 Johnson, W. and Mellor, P.B. Engineering Plasticity. Van Nostrand, Reinbold, 1973.

136 Kamenomostskaja (Kamin), S.L. On Stefan's Problem. (in Russian) Mat. Sb. 53 (1961), 489-514.

137 Kamin, S. Similarity Solutions in Partial Differential Equations. Arch. Rat. Mech. Anal. 60 (1976), 171-179.

138 Kinderlehrer, D. Variational Inequalities and Free Boundary Problems. Bull. Am. Math. Soc. 84, (1978), 7-26.

139 Kinderlehrer, D. The Smoothness of the Solution of the Boundary Obstacle Problem. J. Math. Pur. et Appl. 60, (1981), 193-212.

140 Kinderlehrer, D. and Stampacchia, G. An Introduction to Variational Inequalities and their Applications. Academic Press, New York, 1980.

141 Knerr, B.F. The Porous Medium Equation in One Dimension. Trans. Am. Math. Soc. 234 (1977), 381-415.

142 Krasovskii, Yu. P. On the Theory of Steady-state Waves of Large Amplitude. U.S.S.R. Comp. Math. & Math. Phys. 1 (1961), 996.

143 Landau, H.G. Heat Conduction in a Melting Solid. Q. J. Appl. Math. 8 (1950), 81-94.

144 Langer, J.S. Instabilities and Pattern Formation in Crystal Growth. Rev. Mod. Phys. 52 (1980), 1-28.

145 Lax, P.D. Weak Solutions of Nonlinear Hyperbolic Equations and their Numerical Computation. Comm. Pure & Appl. Math. 7 (1954), 159-193.

146 Lax, P.D. Hyperbolic Systems of Conservation Laws II. Comm. Pure Appl. Math. 10 (1957), 537-566.

147 Lax, P.D. and Wendroff, B. Difference Schemes for Hyperbolic Equations with High Order of Accuracy. Comm. Pure Appl. Math. 17, (1964), 381-398.

148 Lewy, H. and Stampacchia, G. On the Regularity of the Solution of a Variational Inequality. Comm. Pure App. Math. 20 (1967), 439-519.

149 Lions, J.L. Quelques méthodes de résolution des problèmes aux limites non linèaires. Dunod, Paris, 1969.

150 Lions, J.L. and Stampacchia, G. Variational Inequalities. Comm. Pure Appl. Math. 20 (1967), 493-519.

151 Longworth, D. A Numerical Method to Determine the Temperature Distribution Around a Moving Weld Pool. in [180], 54-61.

152 Luke, J.C. A Variational Principle for a Fluid with a Free Surface. J. Fluid Mech. 27 (1967), 395-398.

153 LULEA Proceedings Internat. Symp. Frost Action in Soils, Lulea (1977).

154 McGeough, J.A. and Rasmussen, H. On the Derivation of the Quasi-Steady Model in Electrochemical Machining. J. Inst. Maths. Applics. 13 (1974), 13-21.

155 McLean, J.W. and Saffman, P.G. The Effect of Surface Tension on the Shape of Fingers in a Hele-Shaw Cell. J. Fluid Mech. 102 (1981), 455-469.

156 Magenes, E. Topics .n Parabolic Equations: Some Typical Free Boundary Problems. in Boundary Value Problems for Linear Partial Differential Equations. Proc. NATO Adv. Stud. Inst. (1976), Garnier, H.G. and Reidel, D., 239-312.

157 Magenes, E. (ed.) Free Boundary Problems. (2 vols.) Proceedings Pavia Bimestre (1979), 1st. Naz. Alta Matem. Francesco Severi, Rome (1980).

158 Mangasarian, O. Solution of Symmetric Linear Complementarity Problems by Iterative Methods. J. Opt. Th. Applics. 27 (1977), 465-485.

159 Mason, C.J. Numerical Methods for Free Boundary Problems. Ph.D. Thesis, Imperial College, London (1981).

160 Matkowsky, B.J., and Sivashinsky, G.I. Propagation of a Pulsating Reaction Front in Solid Fuel Combustion. SIAM J. Appl. Maths. 35 (1978), 465-478.

161 Meyer, G.H. The Numerical Solution of Quasilinear Elliptic Equations, in 'Numerical Solution of Systems of Nonlinear Algebraic Equations. ed. Byrne, G. & Hall, G. Acad. Press, 1973, 27-61.

162 Meyer, G.H. Multidimensional Stefan Problems. SIAM J. Numer. Anal. 10 (1973), 522-538.

163 Meyer, G.H. One-dimensional Parabolic Free Boundary Problems. SIAM Review 19 (1977), 17-34.

164 Meyer, G.H. The Numerical Solution of Multidimensional Stefan Problems - A Survey in [249], 73-90.

165 Moore, D.W. The Equation of Motion of a Vortex Layer of Small Thickness. Stud. Appl. Math. 58 (1978), 119-140.

166 Moré, J. Coercivity Conditions in Nonlinear Complementarity Problems. SIAM Rev. 16 (1974), 1-16.

167 Mosco, U. and Strang, G. One-sided Approximation and Variational Inequalities. Bull. Am. Math. Soc. 80, (1974), 308-312.

168 Muehlbauer, J.C. and Sunderland, J.E. Heat Conduction with Freezing or Melting. App. Mech. Rev., 18 (1965), 951-959.

169 Muskat, M. The Flow of Homogeneous Fluids through Porous Media. McGraw-Hill, New York, 1937.

170 Necas, J. Les Methodes Directes dans la Theorie des Équations aux Derivées Partielles. Masson, Paris, 1967.

171 Neuman, S.P. Theoretical Derivation of Darcy's Law. Acta Mech. 25 (1977), 153-170.

172 Nilson, R.H. and Tsuei, Y.G. Free Boundary Problem for the Laplace Equation with Application to ECM Tool Design. Trans. ASME (J. Appl. Mech.) March 1976, 54-58.

173 Niezgódka, M. and Pawlow, I. A Generalised Stefan Problem in Several Space Variables. App. Math. Opt. (1981) (in press).

174 Nitsche, J. L_∞ Convergence of Finite Element Approximations. Mathematical aspects of finite element methods. Springer Lec. Notes No. 606 (1977), 261-274.

175 Nitsche, J. A Finite Element Method for Parabolic Free Boundary Problems. in [157], vol. 1, 277-318.

176 Nye, J.F. Theory of Regelation. Phil. Mag. 16 (1967), 1249-1266.

177 Ockendon, J.R. Techniques of Analysis. in [180], 138-149.

178 Ockendon, J.R. Numerical and Analytical Solutions. in [249], 129-146.

179 Ockendon, J.R. Linear and Nonlinear Stability of a Class of Moving Boundary Problems. in [157], Vol. II, 443-478.

180 Ockendon, J.R. and Hodgkins, W.R. (eds.) Moving Boundary Problems in Heat Flow and Diffusion. Oxford Univ. Press, Oxford, 1975.

181 Oden, J.T. and Kikuchi, N. Theory of Variational Inequalities with Applications to Problems of Flow Through Porous Media. Int. J. Eng. Sci. 18 (1980), 1173-1284.

182 Oleinik, O.A. Discontinuous Solutions of Nonlinear Differential Equations. Uspekhi Mat. Nauk. 12, (1957), 3-73.

183 Oleinik, O.A. A Method of Solution of the General Stefan Problem. Sov. Math. Dokl., 1 (1960), 1350-1354.

184 Oleinik, O.A., Kalashnikov, A.V. and Shzou Yui-Lin. The Cauchy Problem and Boundary Problems for Equations of the Type of Unsteady Filtration. Izv. Ahad. Nauk S.S.S.R. Ser. Mat. 22 (1958), 667-704.

185 Ortega, J.M. and Rheinboldt, W.C. Iterative Solution of Nonlinear Equations in Several Variables. Academic Press, New York, 1970.

186 Pamplin, B.R. (ed.) Crystal Growth. Pergamon, Oxford, 1975.

187 Panagiotopoulos, P.D. A Variational Inequality Approach to the Friction Problem of Structures with Convex Strain Energy Density and Application to the Frictional Unilateral Contact Problem. J. Struct. Mech. 6 (1978), 303-318.

188 Pawlow, I. and Visintin, A. Generalised Stefan-type Problems Involving Additional Nonlinearities. in [105].

189 Pearson, J.R.A. The Instability of Uniform Viscous Flow under Rollers and Spreaders. J. Fluid Mech., 7 (1960), 481-500.

190 Peel, D.A. Some Moving Boundary Problems in the Steel Industry. in [180], 5-18.

191 Pickering, W.M. and Burley, D.M. The Oscillations of a Simple Castor. Bull. Inst. Maths. Applics. 13 (1977), 47-50.

192 Pitts, E. Penetration of Fluid into a Hele-Shaw Cell. J. Fluid Mech. 97, (1980), 53-74.

193 Polubarinova-Kochina, P.Y. Theory of Ground Water Movement. Princeton, 1962.

194 Prager, W. and Hodge, P.G. Theory of Perfectly Plastic Solids, Wiley, New York, 1951.

195 Primicerio, M. Problemi di diffusione a frontiera libera. Boll. U.M.I. (5) 17-A (1980), 1053.

196 Quarteroni, A. and Visintin, A. On the Numerical Solution of a Nonlinear Variational Equation Related to a Filtration Problem. Boll. U.M.I. (5) 17-B (1980), 204-231.

197 Raviart, P.A. Sur la Résolution et l'Approximation de Certaines Équations Paraboliques Nonlinéaires Dégénerées. Arch. Rat. Mech. Anal. 25 (1967), 64-80.

198 Raviart, P.A. Sur l'Approximation de Certain Équations d'Évolution Linéaires et Non-linéaires. J. Math. Pur. et Appl. 46 (1967), 11-183.

199 Richardson, S. Hele-Shaw Flow with a Free Boundary produced by the Injection of a Fluid into a Narrow Channel. J. Fluid Mech. 56 (1972), 609-618.

200 Richtmyer, R.D. and Morton, K.W. Difference Methods for Initial Value Problems. Wiley, Interscience, New York, 1967.

201 Rinzel, J. and Keller, J.B. Traveling Wave Solutions of a Nerve Conduction Equation. Biophysical J. 13 (1973), 1313-1337.

202 Rivard, J.B. and Scott, M.R. The Two-Phase, Three-Zone Melting Problem - Response to Pulse Heating. App. Math. Comp. 2 (1976), 173-187.

203 Rogers, J.C.W. A Free Boundary Problem as Diffusion with Nonlinear Absorption. J. Inst. Maths. Applics. 20 (1977), 261-268.

204 Rogers, J.C.W. The Formulation of Selected Free Boundary Problems as Conservation Laws. in [157], 319-331.

205 Rogers, J.C.W. Relation of the One-Phase Stefan Problem to the Seepage of Liquids and Electrochemical Machining. in [157], 333-382.

206 Rohde, S.M. and McAllister, G.T. A Variational Formulation for a Class of Free Boundary Problems arising in Hydrodynamic Lubrication. Int. J. Eng. Sci. 13 (1975), 841-850.

207 Rose, M. A Method for Calculating Solutions of Parabolic Equations with a Free Boundary. Math. Comp. 14 (1960), 249-256.

208 Rowson, D.M. An Analysis of Stick-Slip Motion. Wear, 31, (1975), 213-218.

209 Rubinstein, L.I. The Stefan Problem. Trans. Math. Mon., Amer. Math. Soc., Providence, 1971.

210 Rubinstein, L.I. The Stefan Problem: Comments on its Present State. J. Inst. Maths. Applics. 24 (1979), 259-278.

211 Rubinstein, L. Global Stability of the Neumann Solution of the Two-Phase Stefan Problem. J. Inst. Maths. Applics. 1982, (in press).

212 Sacher, R.S. On the Solution of Large, Structured Linear Complementarity Problems: II. Tech. Rep. 73-5, Stanford Univ. Dept. Operations Research, 1973.

213 Saffman, P.G. and Taylor, G.I. The Penetration of Fluid into a Porous Medium or Hele-Shaw cell. Proc. Roy. Soc. A 245 (1958), 312-329.

214 Savage, M.D. Cavitation in Lubrication. J. Fluid Mech. 80 (1977), 743-755.

215 Scarpini, F. and Vivaldi, M.A. Evaluation de l'Erreur d'Approximation pour une Inequation Parabolique Relative Aux Convexes Dépendent du Temps. Appl. Math. Opt. 4 (1978), 121-138.

216 Schatz, A. Free Boundary Problems of Stefan Type with Prescribed Flux. J. Math. Anal. App. 28 (1969), 569-580.

217 Shamsundar, N. and Sparrow, E.M. Analysis of Multidimensional Conduction Phase Change via the Enthalpy Model. Trans. A.S.M.E., J. Heat Transfer, 1975 (Aug.) 333-340.

218 Shercliff, J.A. Seepage Flow in Unconfined Aquifers. J. Fluid Mech. 71 (1975), 181-192.

219 Sherman, B. A General One-Phase Stefan Problem. Quart. J. Appl. Math. 27 (1970), 427-439.

220 Showalter, R.E. Hilbert Space Methods for Partial Differential Equations. Pitman, London, 1977.

221 Siegel, R. Analysis of Solidification Interface Shape during Continuous Casting of a Slab. Int. J. Heat Mass Transf. 21 (1978), 1421-1430.

222 Sobolev, S.L. Applications of Functional Analysis in Mathematical Physics. Trans. Math. Mon. 7, Amer. Math. Soc., 1963.

223 Southwell, R.V. Relaxation Methods. Oxford Univ. Press, Oxford, 1946.

224 Spence, D.A. An Eigenvalue Problem for Elastic Contact with Finite Friction. Proc. Camb. Phil. Soc. 73 (1973), 249-268.

225 Spence, D.A. The Hertz Contact Problem with Finite Friction. J. Elast. 5 (1975), 297-319.

226 Stampacchia, G. Formes Bilineaires Coercitives sur les Ensembles Convexes. C.R. Acad. Sci. Paris 258 (1964), 4413-4416.

227 Stewartson, K. and Waechter, R.T. On Stefan's Problem for Spheres. Proc. Roy. Soc. A 348 (1976), 415-426.

228 Stoker, J.J. Water Waves. Interscience, New York, 1957.

229 Stuart, C. An Introduction to the Stefan Problem. Lecture Notes Ecole Polytechnique Fédérale de Lausanne, 1980.

230 Stuart, C.A. and Toland, J. A Variational Method for Boundary Value Problems with Discontinuous Nonlinearities. J. Lond. Math. Soc. 21 (1980), 319-328.

231 Tacke, K-H. Macrosegregation in strand cast steel: Computation of Concentration Profiles with a Diffusion Model. Arch. Eisenhüttenwer. 52 (1981), 15-20.

232 Taubert, K. Differenzverfahren für Schwingungen mit Trockener und Zäher Reibung und für Regelungssysteme. Num. Math. 26 (1976), 379-395.

233 Tayler, A.B. The Mathematical Formulation of Stefan Problems. in [180], 120-137.

234 Tayler, A.B. Fluid Flow between a Roller and Absorbent Compressible Paper. Quart. J. Mech. Appl. Math. 31 (1978), 481-495.

235 Temam, R. A Nonlinear Eigenvalue Problem: The Shape at Equilibrium of a Confined Plasma. Arch. Rat. Mech. Anal. 60 (1975), 51-73.

236 Torrelli, A. Existance and Uniqueness of the Solution for a Nonsteady Free Boundary Problem. Boll. U.M.I. (5) 14-B (1977), 423-466.

237 Torelli, A. On the Study of the Free Boundary for an Elliptic Fourth Order Variational Inequality. in [157] Vol. II, 575-582.

238 Turland, B.D. A General Formulation of the Isotherm Migration Method for Reactor Accident Analysis. Rep. CLM-P582, 1979, U.K.A.E.A. Culham Lab., Abingdon, Oxfordshire.

239 Turland, B.D. and Peckover, R.S. The Stability of Planar Melting Fronts in Two-Phase Thermal Stefan Problems. J. Inst. Maths. Applics. 25 (1980), 1-16.

240 Turner, J.R. The Frictional Unloading Problem on a Linearly Elastic Half-Space. J. Inst. Maths. Applics. 24 (1979), 439-469.

241 Van Moerbeke, P. On Optimal Stopping and Free Boundary Problems. Arch. Rat. Mech. Anal. 60 (1976), 101-148.

242 Varga, R. Matrix Iterative Analysis. Prentice Hall, 1962.

243 Visintin, A. Study of a Free Boundary Filtration Problem by a Nonlinear Variational Equation. Boll. U.M.I. (5) 16-B (1979), 212-237.

244 Visintin, A. Existence Results for Some Free Boundary Filtration Problems. Ann. di Matem. Pur. Appl. CXXIV (1980), 293-320.

245 Voller, V. and Cross, M. Accurate Solutions of Moving Boundary Problems using the Enthalpy Method. Int. J. Heat Mass Transf. 24 (1981), 545-556.

246 Wagner, H. Über Stoss-und Gleit vorgänge au der Oberfläche von Flüssigkeiten. Z. Angew. Math. Mech. 12 (1932), 193-215.

247 Wheeler, J.A. Permafrost Thermal Design for the Trans-Alaska Pipeline. in [249], 267-284.

248 Whitham, G.B. Linear and Nonlinear Waves. Wiley, New York, 1974.

249 Wilson, D.G., Solomon, A.D. and Boggs, P.T. (eds.) Moving Boundary Problems, Academic Press, New York, 1978.

250 Wollkind, D.J. and Segel, L.A. A Nonlinear Stability Analysis of the Freezing of a Dilute Binary Alloy. Phil. Trans. 268A (1970), 351-380.

251 Wooding, R.A. and Morel-Seytoux, N.J. Multiphase Fluid Flow through Porous Media. Ann. Rev. Fluid Mech. 8 (1976), 233-274.

252 Yih, C-S. Stratified Flows. Academic Press, New York, 1980.

253 Zienkiewicz, O.C., Parekh, C.J. and Wills, A.J. The Application of Finite Elements to Heat Conduction Problems Involving Latent Heat. Rock Mechanics 5 (1973), 65-76.

Appendix

In this appendix we will briefly list some definitions and theorems and describe some ideas from functional analysis. This is intended simply as an aide-memoire and we refer to [A1, A2, A3] for expositions of functional analysis and function spaces and to [A4, 140, 170, 220] for descriptions of Sobolev spaces and their application to partial differential equations.

1. CONVERGENCE

We use the notation $u^k \to u$ to mean that the norm $\|u_k - u\| \to 0$ in a normed vector space V. This definition of convergence of a sequence $\{u^k\}$ as $k \to \infty$ is also sometimes called strong convergence.

2. LINEAR FUNCTIONALS

A mapping $\ell(.): V_1 \to V_2$ between two normed vector spaces V_1 and V_2 is continuous if $u^k \to u$ in V_1 implies $\ell(u^k) \to \ell(u)$ in V_2. If $V_2 = \mathbb{R}$, ℓ is called a functional. The space of all continuous (or bounded) *linear* functionals on V is called the *dual space* V^*.

3. WEAK CONVERGENCE

The sequence $\{u^k\}$ converges weakly to u if, for every $\ell \in V^*$, $\ell(u^k) \to \ell(u)$. Weak convergence is denoted by $u^k \rightharpoonup u$, and u is unique.

4. COMPACTNESS

A subset S of a Banach (complete normed) space is called *compact* or *sequentially compact* if and only if every sequence in S contains a convergent subsequence with a limit in S. The set $\{v \in V, \|v\| \le 1\}$ is compact if and only if the vector space V is finite dimensional.

5. RIESZ REPRESENTATION THEOREM (pp. 99, 100)

Let H be a Hilbert space with inner product $(.,.)$. Then, corresponding to every element $\ell \in H^*$, there is a unique $\tilde{\ell} \in H$ such that

$$\ell(u) = (\tilde{\ell},u)$$

for all $u \in H$.

6. WEAK CONVERGENCE IN A HILBERT SPACE

As in 3 above, $\{u^k\}$ is weakly convergent in H if, for every $v \in H$, $(u^k,v) \to (u,v)$.

7. WEAK SEQUENTIAL COMPACTNESS

Every bounded sequence $\{u^k\}$ in a Hilbert space H has a weakly convergent subsequence $\{u^{k'}\}$, i.e. there is a u such that $u^{k'} \rightharpoonup u$.

8. HILBERT SPACE PROJECTION THEOREM (pp. 98, 99)

Let K be a closed convex set of a Hilbert space H. Then for each $u \in H$, there is a unique $v = P_K(u)$ called the projection of u on K such that

$$\|u-v\| = \inf_{w\in K} \|u-w\|$$

and

$$\|P_K(u)-P_K(w)\| \leq \|u-w\| \text{ for } u,w \in H.$$

9. MONOTONE OPERATORS (p. 133)

If D is a subset of the vector space V, $A:D \to V^*$ is said to be a monotone operator if

$$(A(u)-A(v))(u-v) \geq 0 \text{ for all } u,v \in D.$$

The classical solution of a free or moving boundary problem satisfies differential equations away from the unknown boundary and discontinuity conditions of some sort at the unknown boundary. In order to be able to reformulate mathematically such a classical statement as a fixed domain problem, it may be necessary to generalise the idea of classical differentiation to account for derivatives at the unknown boundary. This is why the theory of distributions and generalised functions is important for this book.

10. DISTRIBUTIONS

Here and henceforth let Ω be an open bounded set in $\mathbb{R}^n$ with Lipschitz continuous boundary $\partial\Omega$ so that we can apply Green's theorem. The closure of Ω is $\bar{\Omega} = \Omega \cup \partial\Omega$. We denote by (p. 63)

$$C^k(\Omega) \text{ or } C^k(\bar{\Omega}), \quad k \geq 0$$

the set of functions having continuous derivatives of all orders $\leq k$ in Ω or $\bar{\Omega}$ respectively. The set $C^k(\bar{\Omega})$ is a Banach space with norm

$$\|u\|_{C^k(\bar{\Omega})} = \max_{0\leq|\underset{\sim}{\alpha}|\leq k} \sup_{\underset{\sim}{x}\in\Omega} |D^{\underset{\sim}{\alpha}}u(\underset{\sim}{x})|, \quad \underset{\sim}{\alpha} = (\alpha_1,\ldots,\alpha_n)$$

where $|\underset{\sim}{\alpha}| = \Sigma_{i=1}^n \alpha_i$ and $D^{\underset{\sim}{\alpha}}u = \partial^{|\underset{\sim}{\alpha}|}u/\partial x_1^{\alpha_1}\ldots\partial x_n^{\alpha_n}$.

The support of a function u is the closure of the set on which $u \neq 0$ and we denote by (p. 64)

$$C_0^\infty(\Omega)$$

the set of infinitely differentiable functions with compact support in Ω.

We denote by (p. 67)

$$L^2(\Omega)$$

the Hilbert space consisting of Lebesque square-integrable functions with scalar product $(u,v)_{L^2(\Omega)} = \int_\Omega uvd\underset{\sim}{x}$ and norm

$$|u|_0 = \{(u,u)_{L^2(\Omega)}\}^{\frac{1}{2}}.$$

However, for general $p \geq 1$, we denote by (p. 73)

$$L^p(\Omega)$$

the Banach space of measurable functions with norm

$$\|u\|_{L^p(\Omega)} = \{\int_\Omega |u|^p dx\}^{1/p} < \infty.$$

$L^\infty(\Omega)$ is the Banach space of bounded functions on Ω and

$$\|u\|_{L^\infty(\Omega)} = \inf\{k: |u(\underset{\sim}{x})| \le k \text{ a.e. in } \Omega\}.$$

We are now in a position to define a distribution. We consider the vector space $V(\Omega)$ which consists of the elements of $C_0^\infty(\Omega)$ and convert it into a topological vector space by defining convergence in $V(\Omega)$ by $\phi_n(\underset{\sim}{x}) \to \phi(\underset{\sim}{x})$ if

(i) ϕ_n has compact support in $V(\Omega)$

(ii) $D^{\underset{\sim}{\alpha}}\phi_n \to D^{\underset{\sim}{\alpha}}\phi$ uniformly for all $\underset{\sim}{\alpha}$.

This definition allows us to define not only linear functionals on $V(\Omega)$ but also *continuous* linear functionals and these are called *distributions*. Thus the value of a distribution T on a function $\phi \in C_0^\infty(\Omega)$ is denoted by $T(\phi)$ or $\langle T,\phi\rangle$. It is easy to see that a function $f \in L^1(\Omega)$ defines a distribution whose value on ϕ is

$$\int_\Omega f\phi \, d\underset{\sim}{x} \qquad (*)$$

but not all distributions are of this form. For example, suppose $\underset{\sim}{0} \in \Omega$ and consider the distribution whose value on ϕ is $\phi(\underset{\sim}{0})$. There is no classical function $\delta(\underset{\sim}{x})$ such that $\int_\Omega \delta\phi d\underset{\sim}{x} = \phi(\underset{\sim}{0})$ for all $\phi \in C_0^\infty(\Omega)$ but the Dirac delta function is the distribution which has this property.

The definition of the derivative of a distribution is motivated by the observation that if $f \in C^{|\underset{\sim}{\alpha}|}(\Omega)$, integration by parts $|\underset{\sim}{\alpha}|$ times gives

$$\int_\Omega \phi(\underset{\sim}{x}) D^{\underset{\sim}{\alpha}} f(\underset{\sim}{x}) \, d\underset{\sim}{x} = (-1)^{|\underset{\sim}{\alpha}|} \int_\Omega f(\underset{\sim}{x}) D^{\underset{\sim}{\alpha}}\phi(\underset{\sim}{x}) \, d\underset{\sim}{x}.$$

Hence, bearing in mind (*) above, we define $D^{\underset{\sim}{\alpha}}T$ by

$$(D^{\underset{\sim}{\alpha}}T)\phi = (-1)^{|\underset{\sim}{\alpha}|} T(D^{\underset{\sim}{\alpha}}\phi);$$

$D^{\underset{\sim}{\alpha}}T$ is a continuous linear functional and hence a distribution. Now if

$u \in L^1(\Omega)$ and there exists a $g_{\underset{\sim}{\alpha}} \in L^1(\Omega)$ such that

$$\int_\Omega g_{\underset{\sim}{\alpha}}(\underset{\sim}{x})\, \phi(\underset{\sim}{x})\, d\underset{\sim}{x} = (-1)^{|\underset{\sim}{\alpha}|} \int u(\underset{\sim}{x})\, D^{\underset{\sim}{\alpha}}\phi(\underset{\sim}{x})\, d\underset{\sim}{x}$$

for all $\phi \in V(\Omega)$, then $g_{\underset{\sim}{\alpha}}$ is called the weak or distributional or generalised derivative of u and is denoted by $D^{\underset{\sim}{\alpha}}u$. $D^{\underset{\sim}{\alpha}}u$ is unique up to a set of measure zero and if u has a classical derivative of order $|\underset{\sim}{\alpha}|$ it is also the weak derivative. This definition is used in Chapters III, IV, VI.

The function spaces $C^k(\Omega)$ which we have been using so far in the appendix have the undesirable feature that they are not closed with respect to the norm $\|v\|_{k,p} = \{ \sum_{0 \le |\underset{\sim}{\alpha}| \le k} \|D^{\underset{\sim}{\alpha}}v\|^p_{L^p(\Omega)} \}^{1/p}$. It is to "fill the gaps" left by the limits of sequences which converge under this norm that we are led to consider Sobolev spaces.

11. SOBOLEV SPACES

If k is an integer and $1 \le p \le \infty$, the Sobolev space $W^{k,p}(\Omega)$ is defined by (p. 73)

$$W^{k,p}(\Omega) = \{u \in L^p(\Omega) : D^{\underset{\sim}{\alpha}}u \in L^p(\Omega) \text{ for } 0 \le |\underset{\sim}{\alpha}| \le k\}$$

where $D^{\underset{\sim}{\alpha}}u$ is the weak derivative defined above. It is shown in [A1] that $W^{k,p}(\Omega)$ is a Banach space, with norm $\|u\|_{k,p} = \{ \sum_{0 \le |\underset{\sim}{\alpha}| \le k} \|D^{\underset{\sim}{\alpha}}u\|^p_{L^p(\Omega)} \}^{1/p}$ and that it is the completion of the space $\{u \in C^k(\Omega) : \|u\|_{k,p} < \infty\}$ with respect to the norm $\|\cdot\|_{k,p}$.

On p. 71 et seq we have used the particular notation

$$H^k(\Omega) = W^{k,2}(\Omega).$$

The Sobolev space $H^1(\Omega)$ is a Hilbert space, with inner product

$$(u,v)_{H^1(\Omega)} = \int_\Omega (\nabla u . \nabla v + uv)\, d\underset{\sim}{x}$$

Thus, using the standard notation (p. 71 et seq.)

$$|u|_0^2 = \int_\Omega |u|^2 dx, \qquad |u|_1^2 = \int_\Omega |\nabla u|^2 dx,$$

$$\|u\|_{1,2}^2 = |u|_0^2 + |u|_1^2 .$$

One other important idea we have used concerns the boundary values of functions defined in Ω.

12. TRACE OPERATORS

Suppose u is defined on $\partial\Omega$ and belongs to $L^2(\partial\Omega)$, the Hilbert space of square integrable functions on $\partial\Omega$ with $\|u\|^2_{L^2(\partial\Omega)} = \int_{\partial\Omega} u^2 d\underset{\sim}{S}$. Then it can be shown, [A1, 170, 220] that, if $u \in C^\infty(\bar{\Omega})$,

$$\|u\|_{L^2(\partial\Omega)} \leq c(\Omega)\|u\|_{1,2}$$

where c is a constant depending on Ω. Hence, since $H^1(\Omega)$ is the closure of $C^\infty(\bar{\Omega})$ with respect to the norm $\|\cdot\|_{1,2}$, there is a continuous linear operator $\mathrm{Tr}: H^1(\Omega) \to L^2(\partial\Omega)$ such that

$$\mathrm{Tr}\, u(\underset{\sim}{x}) = u(\underset{\sim}{x}) \quad \text{for } \underset{\sim}{x} \in \partial\Omega,\ u \in C^\infty(\bar{\Omega}).$$

In fact this result is usually written for $u \in H^1(\Omega)$ when there is no risk of confusion.

Whereas we cannot define boundary values for arbitrary functions in $L^2(\Omega)$, since $\partial\Omega$ has zero measure in $\mathbb{R}^n$, we can now define boundary values for functions in $H^1(\Omega)$. In particular we may write (p. 73 et seq)

$$H_0^1(\Omega) = \{u \in H^1(\Omega) : u = 0 \text{ on } \partial\Omega\}.$$

Functions in $H^1(\Omega)$ are not necessarily continuous, and the following theorem gives sufficient additional conditions for continuity. This theorem has been used to prove continuity for the solution of the obstacle problem in theorem IV.4, p. 103.

13. SOBOLEV'S EMBEDDING THEOREM

If $u \in W^{1,p}(\Omega)$ for any $p > n$, then u is almost everywhere equal to a unique function in $C(\bar{\Omega})$. Hence, if $\Omega \in \mathbb{R}^2$, then $u \in H^2(\Omega) \Rightarrow u \in C(\bar{\Omega})$

and $u \in W^{2,p}(\Omega)$, $p > 2 \Rightarrow u \in C^1(\bar{\Omega})$.

---

14. GREENS FORMULA

For $u,v \in H^1(\Omega)$, Greens formula

$$\int_\Omega u \frac{\partial v}{\partial x_i} d\underset{\sim}{x} = - \int v \frac{\partial u}{\partial x_i} d\underset{\sim}{x} + \int_{\partial\Omega} uv \, \nu_i d\underset{\sim}{S}$$

for any $i \in [1,n]$.

Two integrable functions in $L^1(\Omega)$ are said to be equal if they only differ on a set of measure zero. Also a function u is said to be non-negative if $u \geq 0$ except on a set of measure zero. However, in considering the solutions of free and moving boundary problems in Sobolev spaces, we may only be able to define the sets on either side of the known boundary if we have the following stronger definition of positivity, [140].

15. POSITIVITY IN $H^1(\Omega)$

Let $u \in H^1(\Omega)$. We say $u(\underset{\sim}{x}) > 0$ for $\underset{\sim}{x} \in \Omega$ in the sense of $H^1(\Omega)$ if there exists an open set B containing $\underset{\sim}{x}$ and a $\phi \in W^{1,\infty}(B)$ with $\phi \geq 0$ and $\phi(\underset{\sim}{x}) > 0$ such that $u - \phi \geq 0$ a.e. in B. The set $\{\underset{\sim}{x} \in \Omega : u(\underset{\sim}{x}) > 0\}$ is open.

The following inequality in $H_0^1(\Omega)$ has been used in various contexts.

16. POINCARÉ INEQUALITY

There is a constant $c(\Omega)$ such that, for all $u \in H_0^1(\Omega)$,

$$|u|_0 \leq c(\Omega)\,|u|_1$$

This result implies that $|.|_1$ is a norm equivalent to $\|.\|_{1,2}$ on $H_0^1(\Omega)$, since $|u|_1^2 \leq \|u\|_{1,2}^2 \leq \{1 + c^2(\Omega)\}|u|_1^2$ for all $u \in H_0^1(\Omega)$. We also note that $c^2(\Omega)$ is bounded above by the smallest eigenvalue of $-\nabla^2$ on Ω with Dirichlet boundary conditions, since this eigenvalue is itself less than or equal to $\int_\Omega |\nabla u|^2\,dx/\int |u|^2 dx$.

We have often required the following theorem to prove convergence in mean of certain numerical approximations.

17. H^1 COMPACTNESS THEOREM, [220].

A bounded sequence in $H^1(\Omega)$ has a subsequence which is convergent in $L^2(\Omega)$.

---

18. TIME-DEPENDENT PROBLEMS

For time-dependent problems, we need to consider elements of normed vector spaces which depend on a parameter $t \in (0,T)$.

If X is a Banach space, we denote by (p. 82)

$$L^p(0,T;X), \quad 1 \leq p < \infty$$

the space of all measurable functions $x(t)$ taking values in X for almost all $t \in (0,T)$, such that $\|x\|_X^{p'}$ is Lebesque integrable over $(0,T)$. $L^p(0,T;X)$ is a Banach space with norm given by

$$\int_0^T \|x\|_X^p dt$$

and, in the notation of Chapter III,

$$L^2(0,T;L^2(\Omega)) = L^2(\Omega_T).$$

Similarly we denote by

$$C[0,T;X]$$

the set of functions from $[0,T]$ into X which are continuous. When H is a

Hilbert space, a similar definition applies to $L^2[0,T;H]$, which is a Hilbert space with scalar product $\int_0^T (.,.)\ dt$.

A result similar to Sobolev's embedding theorem is that if $u \in L^2(0,T;H)$ and $\partial u/\partial t \in L^2(0,T;H)$, then u is equal to a unique function in $C[0,T;H]$ a.e. in $(0,T)$.

REFERENCES

A1 Adams, R.A. Sobolev Spaces. Academic Press, New York, 1975.

A2 Aubin, J-P. Applied Functional Analysis. Wiley, New York, 1979.

A3 Hutson, V. and Pym, J.S. Applications of Functional Analysis and Operator Theory. Academic Press, London, 1980.

A4 Lions, J.L. and Magenes, E. Non-homogeneous Boundary Value Problems and Applications, Vol. 1. Springer-Verlag, Berlin, 1972.

A5 Milne, R.D. Applied Functional Analysis; An Introductory Treatment. Pitman, London, 1980.

Index